What's that
FLOWER?

What's that
FLOWER?

Dudley Edmondson

DK

**LONDON, NEW YORK, MUNICH,
MELBOURNE, AND DELHI**

DK LONDON
Project Art Editor Francis Wong
Project Editor David Summers
Editorial Assistant Lili Bryant
US Senior Editor Rebecca Warren
US Editor Jill Hamilton
Pre-production Producer
Nikoleta Parasaki
Producer Alice Sykes
Jacket Designer Mark Cavanagh
CTS Sonia Charbonnier
Managing Art Editor Michelle Baxter
Managing Editor Angeles Gavira
Publisher Sarah Larter
Art Director Philip Ormerod
Associate Publishing Director Liz Wheeler
Publishing Director Jonathan Metcalf

DK DELHI
Deputy Managing Art Editor
Sudakshina Basu
Managing Editor Rohan Sinha
Senior Art Editor Anuj Sharma
Senior Editor Vineetha Mokkil
Designers Kanika Mittal,
Divya PR, Upasana Sharma
Editors Suefa Lee, Rupa Rao
DTP Designer Shanker Prasad
Mohammed Usman, Arvind Kumar
DTP Manager/CTS Balwant Singh
Production Manager Pankaj Sharma
Picture Researcher Aditya Katyal

First published in the United States in 2013 by
DK Publishing
375 Hudson Street, New York,
New York 10014

13 14 15 16 10 9 8 7 6 5 4 3 2 1
001 – 191230 – Apr/2013

A catalog record for this book
is available from the Library of Congress
ISBN 978-1-46540-592-0
DK books are available at special discounts when
purchased in bulk for sales promotions, premiums,
fund-raising, or educational use. For details contact:
DK Publishing Special Markets, 375 Hudson Street,
New York, 10014 or SpecialSales@dk.com
Printed and bound in China by
South China Printing Co. Ltd.

Discover more at
www.dk.com

ABOUT THE AUTHOR

Dudley Edmondson is
a nature photographer,
filmmaker, and author.
He began his journey by
exploring the US to
photograph birds, plants,
and animals. During this
time it became clear to him
that nature photography
was his calling. Over the
years, he has written and
provided photographs for
many books, including DK's
Bird, and lectured at
universities across the US. In
2009, he was a runner-up
for MIT's Knight Science
Journalism Fellowship.
Dudley currently lives and
works in Duluth, Minnesota.

Contents

Introduction 6
Identifying Wildflowers 8

FLOWER PROFILES   15

1 GREEN
FLOWERS   16

2 WHITE
FLOWERS   26

3 YELLOW–ORANGE
FLOWERS   46

4 RED–PINK
FLOWERS   68

5 PURPLE–BLUE
FLOWERS   86

FLOWER GALLERY   104

Scientific Names 118
Glossary 122
Index 124
Acknowledgments 128

Introduction

This book is your key to the wonderful world of wildflowers
in the eastern region of North America. More than 150
species are profiled—from some of the common wildflowers
that spring up on your lawn and around your favorite city, to
the rarer ones found in national parks and farther afield. In
springtime, the lengthening days transform dull landscapes
east of the Mississippi into colorful patches of wildflowers.
A fascinating array of flowers in different shapes, colors, and
sizes bloom here. This book uses a simple approach to help
you identify them easily: in the catalog section (pp.18–103),
flowers are arranged by color and then by size. The flower
gallery (pp.104–117) organizes them according to plant
families. This guide is simple enough to be a useful tool for
beginners, as well as for parents and teachers who are keen
to introduce children to the rich heritage of wildflowers. The
more you explore the natural world, the more you will want
to learn. Let these pages be your guide, as you begin a
journey of discovery.

Dudley Edmondson

Dudley Edmondson

Plant Anatomy

To identify plants, you need a basic working knowledge of their parts—particularly flowers and leaves. Usually, both will be present at flowering time, although a small number of plants bloom before their leaves appear. Fruit and seeds (p.14) are also important clues for identification.

Petal

Leaf margin

Flower stalk

Leaf stalk

Leaf

Stem

DOWNY YELLOW VIOLET

Cluster of flowers

Node

DAME'S ROCKET

Stems, leaves, and flowers

Instead of growing at random, leaves and flowers sprout from points called nodes, which are arranged at fixed intervals along plant stems. In some plants, such as Dame's Rocket, flowers grow in clusters of up to 30 on a stem. In others, such as Downy Yellow Violet, the flowers bloom singly on slender stalks.

Shape and Growth

Plants have characteristic shapes and ways of growing, which can help you identify them even when they are not in bloom. Some form patches or mats, while others grow singly or in scattered groups. Climbers cling to other plants, using these for support as they grow toward the light.

AQUATIC
Bullhead Lily roots in mud at the bottom of ponds and lakes. Its leaves and flowers grow on the water surface.

CREEPING
Common Strawberry spreads by long, overground stems called runners. These runners produce new plants at their nodes.

PATCH-FORMING
Yellow Wood Sorrel spreads by underground stems. Taller plants, such as nettles, often spread this way, forming dense clumps.

COLONIAL
Bloodroot typically grows in scattered groups, or colonies. These can contain a few dozen plants or many hundreds.

CLIMBING
Climbers hang on by tendrils, hooks, or by twining around solid supports. Bittersweet Nightshade clambers through other plants.

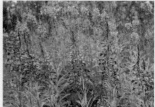

UPRIGHT
Fireweed has tall, upright stems and flower heads containing dozens of blooms. The tall flower heads are easy for insects to spot.

Flower Anatomy

You can identify most flowers easily with the naked eye. The number of petals is often a key feature, as well as their color and shape. Most flowers contain male and female reproductive organs. The male stamens and female stigmas sometimes protrude from the flower.

Petal

Anther at tip of stamen

Stigma

Sepal

WILD GERANIUM

Flower types

The flowers of Wild Geranium and Yellow Wood Sorrel have five petals and five sepals—the flaps that protect the flower as a bud. Purple Coneflower and Bull Thistle flowers are made of many florets—a characteristic feature of the sunflower family (pp.38–39).

Sepal

Flower stalk

YELLOW WOOD SORREL

Disk florets

Ray florets

Ray florets

PURPLE CONEFLOWER

Short, spiny bracts

BULL THISTLE

Flower Shapes

A flower's petals can be separate or they may fuse to form a tube. The simplest flowers are round, but some plants—such as peas and orchids—have complex flowers that are easy to recognize.

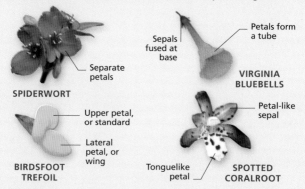

Separate petals

SPIDERWORT

Sepals fused at base

Petals form a tube

VIRGINIA BLUEBELLS

Upper petal, or standard

Lateral petal, or wing

BIRDSFOOT TREFOIL

Petal-like sepal

Tonguelike petal

SPOTTED CORALROOT

Flower Arrangement

Many plants have solitary flowers, with each one growing on a separate stalk. At the other extreme, some of the most eye-catching plants have flowers in large clusters.

Solitary flowers

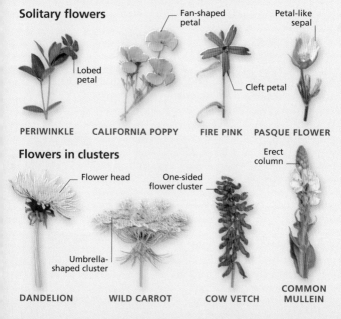

Lobed petal

PERIWINKLE

Fan-shaped petal

CALIFORNIA POPPY

Cleft petal

FIRE PINK

Petal-like sepal

PASQUE FLOWER

Flowers in clusters

Flower head

Umbrella-shaped cluster

DANDELION

WILD CARROT

One-sided flower cluster

COW VETCH

Erect column

COMMON MULLEIN

Leaves

One of the best ways to identify plants is to look
at their leaves. Most plants have one type of leaf.
In some, however, the lower leaves differ in size
and shape from the ones higher up on the stems.
As well as noting shape, look carefully to see
how the leaves are arranged.

Waxy leaf surface

Veins

Leaf anatomy
While leaves typically have a
stalk, some are attached directly
to the stem. Many leaves have
a visible network of spreading
or parallel veins, and an often-
prominent midrib. They can
also have winding tendrils,
which the plant uses to climb.

Midrib
(central vein)

Leaf stalk

COMMON PLANTAIN

Leaf Shapes

Leaves are extremely varied, but most kinds have a single blade,
with or without teeth around the edge. Divided leaves are split into
several leaflets or deep lobes, attached to the same leaf stalk.

TRIANGULAR

ELLIPTICAL

SPOON-SHAPED

HEART-SHAPED

OVAL

KIDNEY-SHAPED

FERNLIKE

OBLONG

ROUNDED, LOBED

Leaf Arrangement

The way that leaves are arranged is as important as their shape. Alternate leaves grow singly along the stem. Opposite leaves are arranged in pairs, with neighboring pairs usually at right angles to each other. Whorled leaves grow in rings.

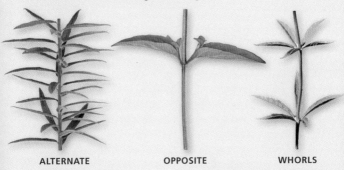

ALTERNATE   OPPOSITE   WHORLS

STAR-SHAPED   ARROW-SHAPED   LANCE-SHAPED

THREE-LOBED   DEEPLY LOBED   THREE-PART

WEDGE-SHAPED   HAND-SHAPED   LADDERLIKE

Fruit and Seeds

Long after their flowers have withered, many plants can be recognized by their fruit. These can be soft and juicy, but others—such as capsules and pods—are hard and dry.

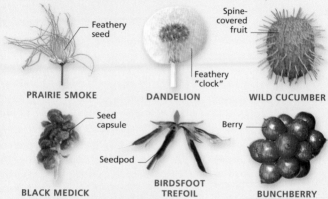

Feathery seed

Spine-covered fruit

Feathery "clock"

PRAIRIE SMOKE

DANDELION

WILD CUCUMBER

Seed capsule

Berry

Seedpod

BLACK MEDICK

BIRDSFOOT TREFOIL

BUNCHBERRY

Plant Habitats

Most plants grow in particular habitats, which helps in identifying the flowers that you find. Some are less fussy—they thrive along paths and roadsides, or in places where the ground has been disturbed. These plants include many common weeds.

GRASSLAND
Colorful vetches are commonly found in meadows and fields.

WOODLAND
Many woodland plants, such as Mayapple, flower early in spring.

SCRUBLAND
California Poppy thrives in open areas, where it grows in dry soil.

WETLAND
Yellow Flag is a typical plant of ponds and slow-flowing water.

ROCKY RIDGES
Columbine—a showy native—grows on rocky outcrops and slopes.

WATERSIDE
Marsh Marigold has fleshy leaves, like many plants found by streams.

FLOWER PROFILES

The following pages will help you identify more than 150 of the most common wildflowers found in the eastern region of North America. The profiles are organized first by color, and then by size. When you are looking at a plant, remember that the same species can often vary—flower sizes, in particular, are only a general guide.

16 GREEN FLOWERS
26 WHITE FLOWERS
46 YELLOW–ORANGE FLOWERS
68 RED–PINK FLOWERS
86 PURPLE–BLUE FLOWERS

Flower sizes

Small Less than ⅜ in (1 cm)
Medium ⅜–¾ in (1–2 cm)
Large More than ¾ in (2 cm)

Symbols

↕ Height
↔ Width

1 GREEN FLOWERS

Even though they are often small and unobtrusive, green flowers are some of the most widespread. Many of them grow in distinctive clusters or flower heads. This group includes a number of pervasive common weeds, such as Leafy Spurge and Curly Dock.

Small Flowers

Plants with small green flowers often grow in disturbed soil. They include some common weeds, such as the widespread Stinging Nettle.

◼ STINGING NETTLE

Prickly plant found along roadsides. In bloom from June to September, male and female flowers grow on separate plants. Its four-sided stems are covered with tiny, bristlelike, stinging hairs. The leaves are heart-shaped.

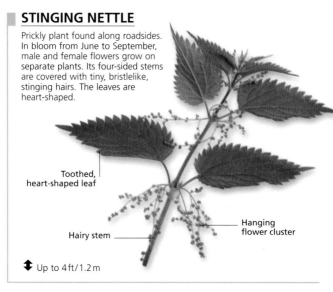

Toothed, heart-shaped leaf

Hairy stem

Hanging flower cluster

⬍ Up to 4 ft / 1.2 m

◼ CURLY DOCK

Robust plant of fields and neglected yards. Its small, brownish or green flowers bloom from June to September. They grow in dense clusters on slender stalks. The leaves, which form a basal rosette and grow alternately on the stem, have distinctly curled edges.

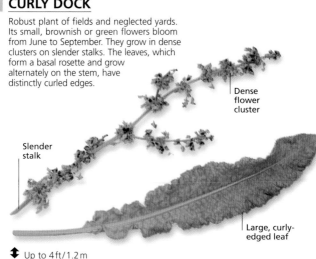

Dense flower cluster

Slender stalk

Large, curly-edged leaf

⬍ Up to 4 ft / 1.2 m

EARLY MEADOW-RUE

Upright plant of wooded areas
and ravines. Its greenish white
flowers bloom early in the year
from April to May. They grow in
small clusters, drooping forward
on long stalks. Male and female
flowers are found on separate
plants. The leaves have three
or four lobes.

Long-stalked
flower cluster

Smooth,
leafy stem

Lobed leaf

✦ Up to 30 in/75 cm

COMMON RAGWEED

Found in waste places and
along roadsides. Its small,
greenish yellow flowers
bloom from July to October.
They grow in elongated
clusters on slender stalks.
Its leaves are deeply divided.
The wind-spread pollen is
a potent allergen for hay
fever sufferers.

Cluster of
tiny flowers

Slender
stalk

Deeply
divided leaf

✦ Up to 5 ft/1.5 m

Medium-sized Flowers

Solomon's Seal is one of the few plants with medium-sized green flowers. It blooms from mid-spring to early summer.

LEAFY SPURGE

Invasive weed of roadsides and waste places. In bloom from May to September, its small, yellowish green flowers grow at the center of two petal-like bracts. The narrow leaves grow along the length of the stems.

Whorl of upper leaves

Small, yellowish green flower

Petal-like bract

Lance-shaped leaf

⬍ Up to 24 in/60 cm

SOLOMON'S SEAL

Graceful plant of moist woodland thickets, in bloom from May to June. The pale green to white flowers appear along the length of the arching stem where each leaf is attached. The stalkless, oval to lance-shaped leaves shade the long row of flowers below.

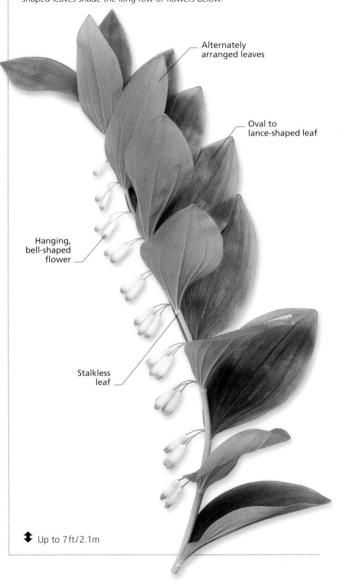

Alternately arranged leaves

Oval to lance-shaped leaf

Hanging, bell-shaped flower

Stalkless leaf

↕ Up to 7 ft / 2.1 m

Large Flowers

Plants with large green flowers, such as Skunk Cabbage, are uncommon, and their odd leaf structures can give them an unusual, almost alienlike, appearance.

JACK-IN-THE-PULPIT

Found on damp forest and woodland floors, and in swamps. Its flowers bloom from April to June. Male and female flowers grow at the base of a long, club-shaped spike, which sits inside a spathe with a curved, rooflike flap. The large, dull green leaves are arranged in whorls of three, atop rigid stems.

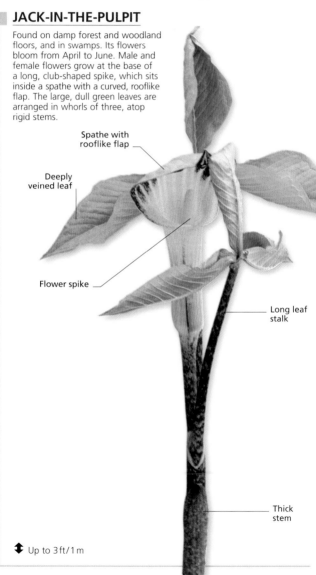

Spathe with rooflike flap

Deeply veined leaf

Flower spike

Long leaf stalk

Thick stem

↕ Up to 3 ft/1 m

SKUNK CABBAGE

Found in marshes, swamps, and moist woodland. Its flowers bloom from February to April, forming a clublike spike that sits inside a curved, greenish brown, spotted spathe. The large, stalked, dark green leaves unfold from a tight roll, which rises directly from the ground. The plant emits a strong, fetid odor when crushed.

Curved, spotted spathe

Clublike spike of flowers

Tight roll of leaves

Plant emerges through moist soil and fallen foliage

↕ Up to 24 in/60 cm

Parsley Family

With more than 2,500 species worldwide—nearly a quarter of which are native to North America—the parsley family includes many wildflowers and aromatic herbs. They are easy to recognize—most have umbrella-shaped flower heads.

Plants in this family are often tall and upright, with hollow and sometimes bristly stems. Their flower heads, called umbels, are usually flat-topped or domed. A single flower head can have dozens of spokelike stalks and hundreds of tightly packed flowers.

Lacy display
Wild Carrot (above) is a typical umbellifer, or member of the parsley family. It sprouts from ground level, growing a tender, orange taproot in its first year and flowering in its second. Its many flower heads attract a wide range of insect pollinators, including bees.

Plant characteristics

Parsley family plants usually have alternate leaves, which are deeply lobed or divided into leaflets. Most have compound flower heads, with a main umbel, which is divided into many smaller ones. Wild Parsnip has yellow flowers, but white is the most common color in the family as a whole.

Branched umbel

Umbel stalk

Oval leaflet

Ridged stem

WILD PARSNIP

Flower anatomy

Plants in the parsley family usually have five-petaled flowers, although in some, the petals are so small that they can be difficult to see. Once the flowers have been pollinated, they produce dry fruit containing two seeds. The fruit's shape varies, and can be a useful clue in identifying a plant.

Five-petaled flower

Protruding stamen

Nectar glands

Ridged case

WILD PARSNIP FLOWERS

WILD PARSNIP FRUIT

2 WHITE FLOWERS

White flowers can be found in bloom throughout the year. Among the earliest to appear are Bloodroot and Sharp-lobed Hepatica, which are commonly found on forest floors. These are followed by a host of other species as spring and summer get underway.

Small Flowers

Plants with small, white flowers often have them grouped in clusters atop tall, upright stems. The low-growing Common Chickweed is an exception.

HOARY ALYSSUM

Tall, erect plant found in waste places and along roadsides. In bloom from June to September, its small, white flowers grow in clusters at the top of the plant. The small, lance-shaped leaves are covered with a fuzzy down.

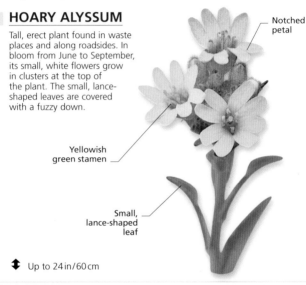

Notched petal

Yellowish green stamen

Small, lance-shaped leaf

✢ Up to 24 in/60 cm

COMMON CHICKWEED

Low-growing plant, widespread in waste places and along roadsides. The tiny white flowers have notched petals, and grow singly or in pairs on erect, hairy stems. They bloom from February to December. The egg-shaped leaves are relatively smooth.

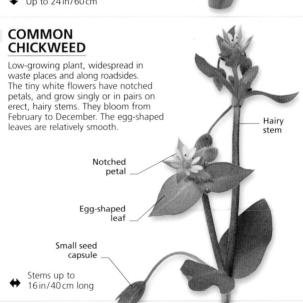

Hairy stem

Notched petal

Egg-shaped leaf

Small seed capsule

↔ Stems up to 16 in/40 cm long

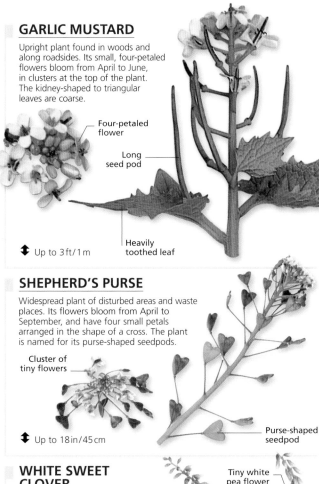

GARLIC MUSTARD

Upright plant found in woods and along roadsides. Its small, four-petaled flowers bloom from April to June, in clusters at the top of the plant. The kidney-shaped to triangular leaves are coarse.

Four-petaled flower

Long seed pod

Heavily toothed leaf

↕ Up to 3 ft / 1 m

SHEPHERD'S PURSE

Widespread plant of disturbed areas and waste places. Its flowers bloom from April to September, and have four small petals arranged in the shape of a cross. The plant is named for its purse-shaped seedpods.

Cluster of tiny flowers

Purse-shaped seedpod

↕ Up to 18 in / 45 cm

WHITE SWEET CLOVER

Tiny white pea flower

Found in fields and along roadsides. Its tiny pea flowers grow on long, erect stalks; they bloom from May to October. Its leaves are divided into three lance-shaped, toothed leaflets. The plant has a vanilla-like fragrance when crushed.

Toothed leaf margin

↕ Up to 8 ft / 2.4 m

»

COW PARSNIP

Tall plant of moist woodland. The flowers bloom from June to August in flat-topped clusters. Their tiny, white petals are often tinted purple. The lobed, toothed, three-part leaves can be up to 12 in (30 cm) wide.

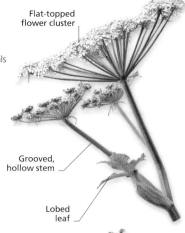

Flat-topped flower cluster

Grooved, hollow stem

Lobed leaf

 Up to 9 ft/2.7 m

POISON HEMLOCK

Poisonous plant that prefers moist soil in waste places and along roadsides. Its delicate-looking, five-petaled flowers bloom from June to August, forming branching, flat-topped clusters. The deeply lobed, divided leaves grow alternately on purple-spotted stems. The plant emits a foul smell when crushed.

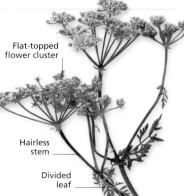

Flat-topped flower cluster

Hairless stem

Divided leaf

Up to 10 ft/3 m

BLACK SNAKEROOT

Common plant of dry open areas and woodland thickets. It produces rounded clusters of tiny, greenish white flowers from May to July. The lobed leaves are divided into three to seven toothed leaflets. They grow at the plant's base and below the stalked flower clusters.

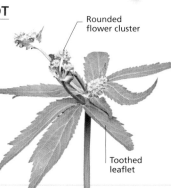

Rounded flower cluster

Toothed leaflet

 Up to 4 ft/1.2 m

WILD CARROT

Found in dry fields and waste places. The flowers bloom from May to October in lacy clusters, forming flat-topped flower heads. The long, fernlike leaves are deeply divided into many leaflets and emit a carrotlike odor when crushed. Also known as Queen Anne's Lace.

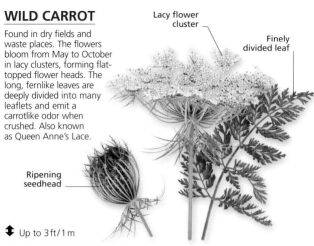

Lacy flower cluster

Finely divided leaf

Ripening seedhead

⬍ Up to 3 ft/1 m

WATER PARSNIP

Aquatic plant of swamps, ponds, and wet meadows. The flowers bloom from July to September in branching clusters. Leaflike bracts grow below the clusters. The leaves are divided into toothed leaflets. The basal leaves are often submerged in the water.

⬍ Up to 6 ft/1.8 m

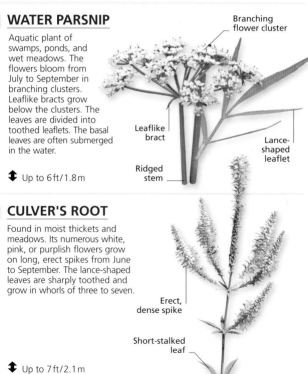

Branching flower cluster

Leaflike bract

Lance-shaped leaflet

Ridged stem

CULVER'S ROOT

Found in moist thickets and meadows. Its numerous white, pink, or purplish flowers grow on long, erect spikes from June to September. The lance-shaped leaves are sharply toothed and grow in whorls of three to seven.

Erect, dense spike

Short-stalked leaf

⬍ Up to 7 ft/2.1 m

»

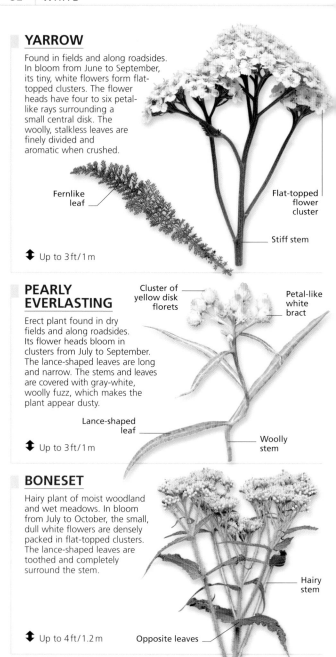

YARROW

Found in fields and along roadsides. In bloom from June to September, its tiny, white flowers form flat-topped clusters. The flower heads have four to six petal-like rays surrounding a small central disk. The woolly, stalkless leaves are finely divided and aromatic when crushed.

Fernlike leaf

Flat-topped flower cluster

Stiff stem

↕ Up to 3 ft / 1 m

PEARLY EVERLASTING

Cluster of yellow disk florets

Petal-like white bract

Erect plant found in dry fields and along roadsides. Its flower heads bloom in clusters from July to September. The lance-shaped leaves are long and narrow. The stems and leaves are covered with gray-white, woolly fuzz, which makes the plant appear dusty.

Lance-shaped leaf

Woolly stem

↕ Up to 3 ft / 1 m

BONESET

Hairy plant of moist woodland and wet meadows. In bloom from July to October, the small, dull white flowers are densely packed in flat-topped clusters. The lance-shaped leaves are toothed and completely surround the stem.

Hairy stem

↕ Up to 4 ft / 1.2 m

Opposite leaves

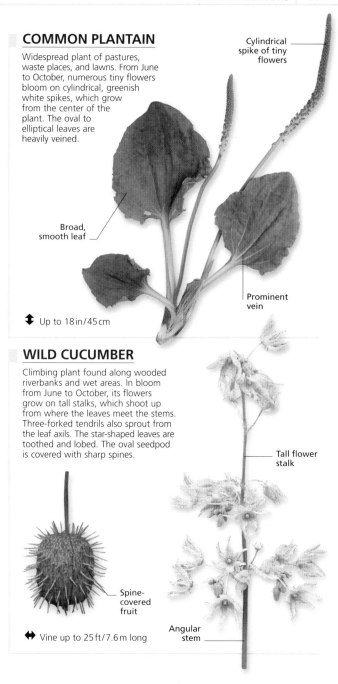

COMMON PLANTAIN

Widespread plant of pastures, waste places, and lawns. From June to October, numerous tiny flowers bloom on cylindrical, greenish white spikes, which grow from the center of the plant. The oval to elliptical leaves are heavily veined.

Cylindrical spike of tiny flowers

Broad, smooth leaf

Prominent vein

✦ Up to 18 in / 45 cm

WILD CUCUMBER

Climbing plant found along wooded riverbanks and wet areas. In bloom from June to October, its flowers grow on tall stalks, which shoot up from where the leaves meet the stems. Three-forked tendrils also sprout from the leaf axils. The star-shaped leaves are toothed and lobed. The oval seedpod is covered with sharp spines.

Tall flower stalk

Spine-covered fruit

↔ Vine up to 25 ft / 7.6 m long

Angular stem

Medium-sized Flowers

Commonly found from spring through to fall, medium-sized white flowers grow in a variety of habitats—from ponds and marshes, to fields and roadsides.

SHARP-LOBED HEPATICA

Found in upland deciduous woods, and on rocky slopes and bluffs. Its white to purplish flowers bloom from March to April. They have 5–11 petal-like sepals surrounding a green central cluster of carpels. The three-lobed leaves have a hairy surface.

Leafy bract

Sharply pointed leaf

↕ Up to 6 in / 15 cm

BLADDER CAMPION

Showy plant of fields and roadsides; in bloom from April to August. The flowers have five deeply-notched petals. The sepals resemble an inflated bladder. The lance-shaped to oval leaves are hairless and typically clasp the stems.

Protruding style and stamens

Hairless stem

Prominently veined sepal

Leaves in opposite pairs

↕ Up to 30 in / 75 cm

BLACK COHOSH

Tall, erect plant of rich woodland areas. Its flower heads, similar to those of Culver's Root (p.31), bloom from June to September. They grow in narrow clusters that form tall spikes, and often fall off after blooming. The large, toothed leaves are often twice divided into threes. The plant has a foul smell.

Hairless stem

Tufted flower head

Round flower bud

↕ Up to 8 ft / 2.4 m

DUTCHMAN'S BREECHES

Found in moist woodland. Its four-petaled, pantaloon-shaped, white flowers bloom from April to May. They grow in clusters on slender stalks. The fernlike basal leaves are deeply divided.

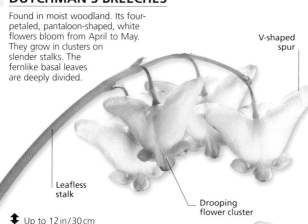

V-shaped spur

Leafless stalk

Drooping flower cluster

✦ Up to 12 in / 30 cm

INDIAN PIPE

Found in shady, rich soil near decaying matter in woodland areas. Also known as Corpse Plant, it is completely white because it contains no chlorophyll. In bloom from June to September, its white or salmon pink flowers have up to five petals. They grow individually on thick, translucent stems, which are covered with scaly bracts.

Translucent stem

Scaly bract

Nodding flower

✦ Up to 10 in / 25 cm

COMMON STRAWBERRY

Creeping plant of open fields and woodland edges. In bloom from April to June, its five-petaled, white flowers have a yellow center with numerous stamens surrounding a small, blunt cone. The toothed leaves grow on short, hairy stalks.

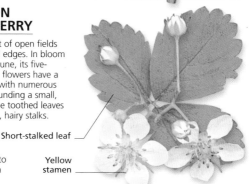

Short-stalked leaf

✦ Stalks up to 6 in / 15 cm

Yellow stamen

»

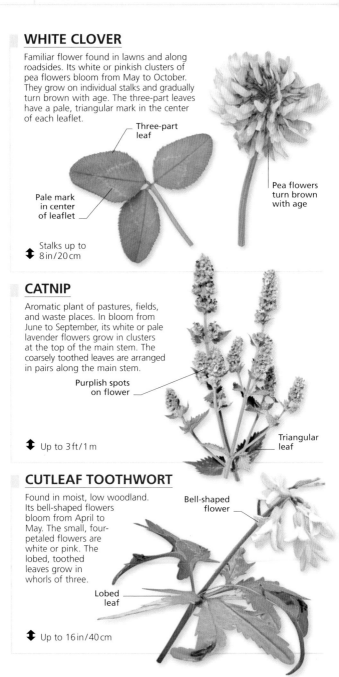

WHITE CLOVER

Familiar flower found in lawns and along roadsides. Its white or pinkish clusters of pea flowers bloom from May to October. They grow on individual stalks and gradually turn brown with age. The three-part leaves have a pale, triangular mark in the center of each leaflet.

Three-part leaf

Pea flowers turn brown with age

Pale mark in center of leaflet

Stalks up to 8 in / 20 cm

CATNIP

Aromatic plant of pastures, fields, and waste places. In bloom from June to September, its white or pale lavender flowers grow in clusters at the top of the main stem. The coarsely toothed leaves are arranged in pairs along the main stem.

Purplish spots on flower

Triangular leaf

Up to 3 ft / 1 m

CUTLEAF TOOTHWORT

Found in moist, low woodland. Its bell-shaped flowers bloom from April to May. The small, four-petaled flowers are white or pink. The lobed, toothed leaves grow in whorls of three.

Bell-shaped flower

Lobed leaf

Up to 16 in / 40 cm

FEVERFEW

Aromatic plant of roadsides and hedgerows. Its flower heads are similar to those of the Oxeye Daisy (p.43), and bloom from June to September. They have a yellow disk at the center, surrounded by short, blunt petal-like rays. The deeply lobed leaves are toothed.

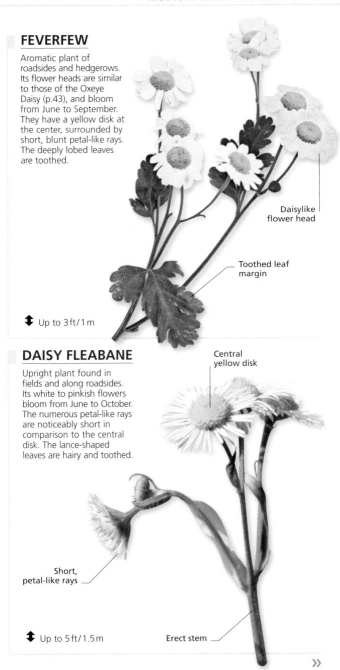

Daisylike flower head

Toothed leaf margin

⬍ Up to 3 ft / 1 m

DAISY FLEABANE

Upright plant found in fields and along roadsides. Its white to pinkish flowers bloom from June to October. The numerous petal-like rays are noticeably short in comparison to the central disk. The lance-shaped leaves are hairy and toothed.

Central yellow disk

Short, petal-like rays

⬍ Up to 5 ft / 1.5 m

Erect stem

»

Sunflower Family

With nearly 25,000 species worldwide, sunflowers and their relatives form one of the largest plant families. They include many wayside and grassland wildflowers.

The plants in this family have tiny flowers, known as florets. These are packed together in composite flower heads, which are easy to mistake for single flowers. Each floret produces just one seed—look out for the hairs or hooks that help them spread.

Winning formula

The Dandelion (above) is one of the most successful sunflower family plants, spreading far and wide with its windblown seeds. As well as dandelions and daisies, this family also includes ragworts, thistles, and knapweeds, and cultivated plants, such as chicory and lettuces.

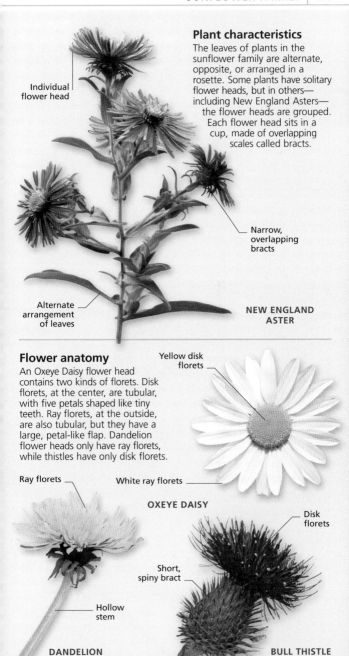

Plant characteristics

The leaves of plants in the sunflower family are alternate, opposite, or arranged in a rosette. Some plants have solitary flower heads, but in others—including New England Asters—the flower heads are grouped. Each flower head sits in a cup, made of overlapping scales called bracts.

Individual flower head

Narrow, overlapping bracts

Alternate arrangement of leaves

NEW ENGLAND ASTER

Flower anatomy

An Oxeye Daisy flower head contains two kinds of florets. Disk florets, at the center, are tubular, with five petals shaped like tiny teeth. Ray florets, at the outside, are also tubular, but they have a large, petal-like flap. Dandelion flower heads only have ray florets, while thistles have only disk florets.

Yellow disk florets

White ray florets

OXEYE DAISY

Ray florets

Hollow stem

DANDELION

Disk florets

Short, spiny bract

BULL THISTLE

WHITE LETTUCE

Tall plant of woodland thickets; also known as Rattlesnake Root. Its bell-shaped, white to purplish flowers bloom from August to September. They have 8–12 downward-hanging petals with prominent, cream-colored stamens. The lower leaves are triangular and can be lobed. The uppermost leaves are lance-shaped.

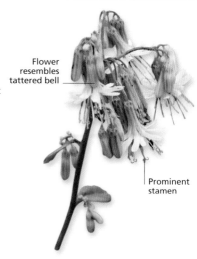

Flower resembles tattered bell

Prominent stamen

⬍ Up to 5 ft / 1.5 m

FLAT-TOP ASTER

Upright plant of moist woodland thickets. In bloom from August to September, its numerous flower heads form flat-topped clusters on rigid stems. They have 10–15 white petal-like rays surrounding a yellow disk.

Lance-shaped leaf

Yellow central disk

⬍ Up to 7 ft / 2.1 m

BROAD-LEAVED ARROWHEAD

Aquatic plant of lakes and swamps. Its flowers bloom from July to September. They have three, rounded to heart-shaped petals surrounding a yellow or green center. The long-stalked leaves are shaped like arrowheads, with two long, backward-projecting lobes.

Heart-shaped petal

Arrow-shaped leaf

⬍ Up to 4 ft / 1.2 m above water

Curved green sepal

Large Flowers

Plants with large, white flowers, such as the distinctive Bunchberry, tend to stand out in any habitat where they thrive.

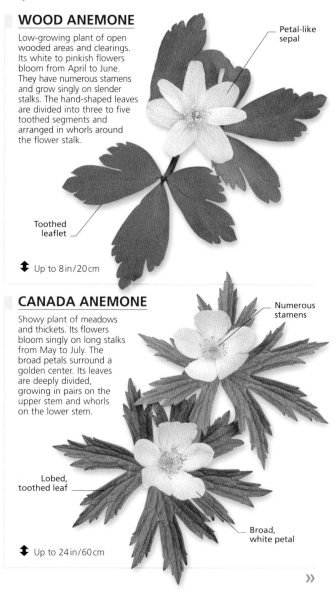

WOOD ANEMONE

Low-growing plant of open wooded areas and clearings. Its white to pinkish flowers bloom from April to June. They have numerous stamens and grow singly on slender stalks. The hand-shaped leaves are divided into three to five toothed segments and arranged in whorls around the flower stalk.

Petal-like sepal

Toothed leaflet

↕ Up to 8 in/20 cm

CANADA ANEMONE

Showy plant of meadows and thickets. Its flowers bloom singly on long stalks from May to July. The broad petals surround a golden center. Its leaves are deeply divided, growing in pairs on the upper stem and whorls on the lower stem.

Numerous stamens

Lobed, toothed leaf

Broad, white petal

↕ Up to 24 in/60 cm

»

MAYWEED

Bushy plant growing in waste places and along roadsides. Its daisylike flower heads bloom from June to October. They have 8–20 petal-like white rays and a dome-shaped yellow center. The fernlike leaves are very finely dissected and give off a foul odor.

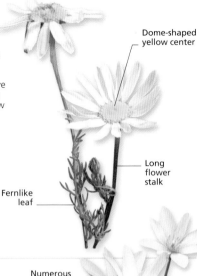

Dome-shaped yellow center

Long flower stalk

Fernlike leaf

✚ Up to 20 in/50 cm

BLOODROOT

Found in rich, moist woodland. In bloom from March to May, the flowers have 8–10 petals and many yellow stamens. Each plant has a hand-shaped basal leaf, often curved around the single flower stalk. The plant gets its name from the red sap inside its underground stems.

Numerous stamens

Bluish green, lobed leaf

✚ Up to 12 in/30 cm

BUNCHBERRY

Low-growing plant of moist, cool, open forested areas. It has four petal-like bracts centered around a yellowish green cluster of tiny flowers, which bloom from May to July. The leaves grow in whorls at the top of stems.

Cluster of tiny flowers

Whorl of oval leaves

White, petal-like bract

✚ Up to 8 in/20 cm

OXEYE DAISY

Widespread, invasive plant found in fields and along roadsides. In bloom from June to August, its flower heads have a yellow central disk surrounded by many white petal-like rays. The heavily lobed leaves grow along the length of the flower stalk.

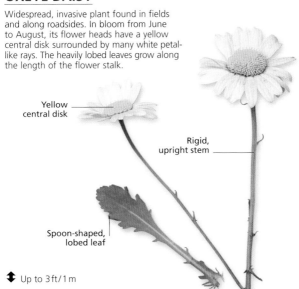

Yellow central disk

Rigid, upright stem

Spoon-shaped, lobed leaf

✸ Up to 3 ft / 1 m

MAYAPPLE

Found in rich woodland and moist clearings. In bloom from March to June, its white flowers have a yellow center surrounded by six to nine petals. They grow singly where two large, deeply lobed leaves join the main stem.

Deeply lobed leaf

Nodding flower

✸ Up to 18 in / 45 cm

»

LARGE-FLOWERED TRILLIUM

Found in rich, moist woods. Its showy white flowers bloom from April to June, on a single stalk at the center of three large, egg- to diamond-shaped leaves. They have three petals and three green sepals.

Narrow, green sepal

Deeply veined leaf

Waxy white petal

⬍ Up to 18 in/45 cm

NODDING TRILLIUM

Perennial plant found in moist, acidic woods and swamps. In bloom from April to July, its three-petaled white flowers nod forward like a pendulum. They have three green sepals. The whorled, diamond-shaped leaves are suspended above the flowers on a single tall stem.

Wavy leaf margin

Whorl of diamond-shaped leaves

Nodding flower

⬍ Up to 24 in/60 cm

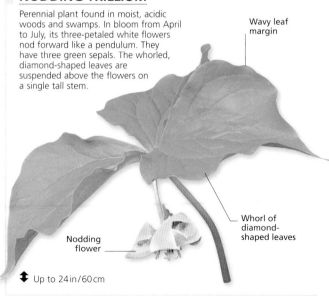

YUCCA

Found in sandy areas and old, dry fields, this striking plant blooms from June to September. Its creamy white, bell-shaped flowers grow on a single tall stem and have wide petals. The rigid, spearlike leaves are slightly concave, with stringlike tendrils growing between them.

Flower bud

Spearlike leaf

Six-petaled flower

Long, thick stem

⬍ Up to 10 ft / 3 m

3 YELLOW–ORANGE FLOWERS

Wherever they grow, plants with yellow or orange flowers are hard to miss—smaller flowers are often grouped in clusters, while larger, showy ones, such as Yellow Flag, bloom singly. Their vibrant colors attract many pollinators like butterflies and honeybees.

Small Flowers

Plants with small yellow or orange flowers, such as Wintercress and Golden Alexanders, can be very conspicuous when the blooms are clustered in flower heads.

▎WINTERCRESS

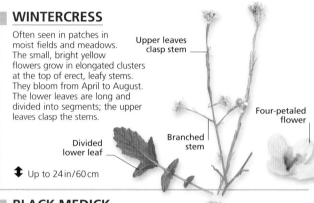

Often seen in patches in moist fields and meadows. The small, bright yellow flowers grow in elongated clusters at the top of erect, leafy stems. They bloom from April to August. The lower leaves are long and divided into segments; the upper leaves clasp the stems.

Upper leaves clasp stem

Four-petaled flower

Branched stem

Divided lower leaf

✦ Up to 24 in/60 cm

▎BLACK MEDICK

Low-growing, sprawling plant found in waste places and along roadsides. The tiny yellow pea flowers grow in dense clusters of up to 50 from March to December. They mature into black seedpods. The divided leaves have three leaflets, each with a tiny spine at the tip.

Long flower stalk

Cluster of tiny flowers

Oval leaflet

Coiled, black seedpods

↔ Stems up to 24 in/60 cm long

▎GOLDEN ALEXANDERS

Perennial plant of meadows and moist woodland thickets. In bloom from April to June, its bright yellow flowers grow in flat-topped clusters. The divided leaves have toothed leaflets that taper to a point at the tip.

Flat-topped flower cluster

Smooth stem

Toothed leaflet margin

✦ Up to 3 ft/1 m

YELLOW SWEET CLOVER

Tall, loosely branched plant of fields and waste places. The tiny yellow pea flowers form cylindrical, spikelike clusters. They bloom from May to October. The narrow, divided leaves have three toothed leaflets that emit a vanilla-like fragrance when crushed.

Flower hangs downward

Egg-shaped leaflet

✤ Up to 5 ft/1.5 m

WILD PARSNIP

Found in moist, open fields and along roadsides. The tiny, greenish yellow flowers bloom in loose, flat-topped, branching clusters from June to July. The divided leaves have 5–15 toothed, oval leaflets. The sap from the stem can cause a serious rash.

Flat, oval seedpod

Yellow stamen

Branched flower cluster

Toothed leaflet margin

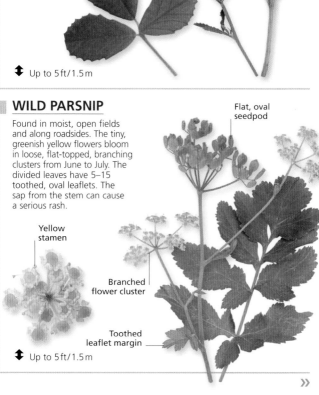

✤ Up to 5 ft/1.5 m

»

BUTTERFLY WEED

Found in fields and dry open soil. In bloom from June to September, its orange flowers have five backward-curving petals and a central crown. The flowers grow in clusters at the top of the plant. The narrow, oblong leaves are alternately arranged on the stems.

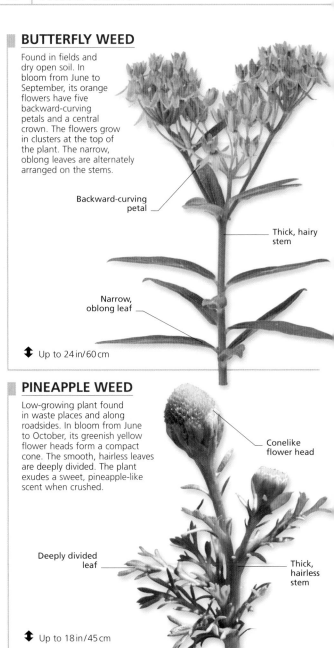

Backward-curving petal

Thick, hairy stem

Narrow, oblong leaf

✤ Up to 24 in/60 cm

PINEAPPLE WEED

Low-growing plant found in waste places and along roadsides. In bloom from June to October, its greenish yellow flower heads form a compact cone. The smooth, hairless leaves are deeply divided. The plant exudes a sweet, pineapple-like scent when crushed.

Conelike flower head

Deeply divided leaf

Thick, hairless stem

✤ Up to 18 in/45 cm

LANCE-LEAVED GOLDENROD

Perennial plant that grows in fields and roadside thickets. Its tiny, yellow flower heads form flat-topped clusters. They bloom from July to October. The lance-shaped leaves have a rough, untoothed margin.

Flat-topped flower cluster

Narrow leaf tapers to a point

Slender, downy stem

Veined leaf

✦ Up to 4 ft / 1.2 m

CANADA GOLDENROD

Tall, showy plant of clearings and roadsides. In bloom from June to September, its tiny, yellow flower heads form dense clusters. They grow on long, flowering branches that arch outward. The lance-shaped leaves are hairy and deeply veined.

Flat-topped flower cluster

Arching branch

Deeply veined leaf

✦ Up to 5 ft / 1.5 m

WILD LETTUCE

Tall plant found in clearings and at the edges of woodland. In bloom from July to September, its dandelionlike, pale yellow flower heads grow in clusters at the top of the plant. The lance-shaped leaves are toothless and deeply lobed.

Dandelion-like flower head

Light green bracts

✦ Up to 10 ft / 3 m

Medium-sized Flowers

Plants with medium-sized yellow or orange flowers are anything but ordinary, with several showy examples, such as Birdsfoot Trefoil, that are easy to spot.

ROUGH-FRUITED CINQUEFOIL

Stout, hairy plant of dry fields and roadsides. In bloom from May to August, its pale yellow flowers have five notched petals. The leaves are divided into five to seven coarsely toothed leaflets.

Lance-shaped leaflet

Yellow-tipped stamen

Hairy stem

✤ Up to 24 in/60 cm

LANCE-LEAVED LOOSESTRIFE

Found in floodplains and damp woodland thickets. In bloom from June to August, its flowers grow on long stalks, facing outward. The minutely toothed oval petals taper to a sharp point. The lance- to egg-shaped leaves are arranged alternately on the stems.

Narrow sepal

Slender stalk

Pointed leaf tip

Red-tinged center

✤ Up to 4 ft/1.2 m

BIRDSFOOT TREFOIL

Found in fields and along roadsides. In bloom from June to September, the yellow to deep orange pea flowers form flat-topped clusters. The leaves are divided into three leaflets, with two leaflike stipules at the base of each stalk. The arrangement of the seedpods resembles a bird's foot.

Leafy bract

Red streaks on petal

Leaflike stipule

✤ Up to 24 in/60 cm

HOP CLOVER

Upright plant of roadsides and waste places. Its pea flowers bloom from June to September, forming roundish oblong flower heads, which resemble dried hops with age. The stalkless leaves grow on smooth stems and are divided into three oblong to lance-shaped leaflets.

Roundish oblong flower head

Prominent veins on leaflet

Smooth stem

✦ Up to 18 in/45 cm

YELLOW WOOD SORREL

Clump-forming plant found in woodland, meadows, and disturbed areas. Its bright yellow flowers bloom singly or in clusters from May to October. The leaves are divided into three heart-shaped leaflets. The upright seed capsules grow on bent stalks.

✦ Up to 16 in/40 cm

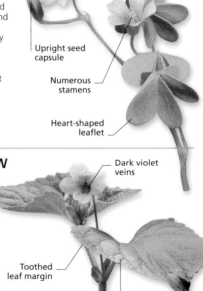

Five-petaled flower

Upright seed capsule

Numerous stamens

Heart-shaped leaflet

DOWNY YELLOW VIOLET

Hairy plant of dry woods that blooms from April to May. The flowers are five-petaled with bearded side-petals and dark violet veins on the lower petals. The heart-shaped, toothed leaves grow from the same stalk as the flowers.

✦ Up to 16 in/40 cm

Dark violet veins

Toothed leaf margin

Broad, deeply veined leaf

»

COMMON TANSY

Patch-forming aromatic plant found in waste places and along roadsides. Its buttonlike, yellow to orange flowers grow from July to September. The long, ladderlike leaves are toothed, and give off a strong foul odor when crushed.

Buttonlike flower head

Long, toothed leaf

⬍ Up to 3 ft / 1 m

GOLDEN RAGWORT

Found in bogs, swamps, and moist woods. In bloom from April to July, its flowers have eight to twelve petal-like rays that surround a central disk. The rays vary in length, which makes the flowers look asymmetrical. The basal leaves are heart-shaped, and lobed leaves grow on the long stems.

Daisylike flower head

Long stem

Lobed stem leaf

↕ Up to 3 ft / 1 m

KING DEVIL

Found in open fields and along roadsides. Its bright yellow flower heads grow in clusters on short stalks, branching from a single hairy main stem. They bloom from May to August. The oblong leaves at the base of the plant are covered with stiff hairs.

Hairy bract

Petal-like ray

Hairy main stem

↕ Up to 3 ft / 1 m

»

Mint Family

Found worldwide, the mint family includes about 7,000 species. Apart from mints, it contains many wildflowers of open places or shaded ground.

Mints and their relatives have a long history of cultivation, with many being planted as culinary herbs. Wild plants are often easy to spot—look out for telltale whorls of two-lipped flowers, which grow on square-sided, upright stems.

Colorful clumps
Many plants in the mint family spread easily using underground stems, called rhizomes, to form dense mats or clumps. Wild Bergamot (above) reproduces using both seeds and rhizomes, forming larger clumps every year.

Plant characteristics

Most members of the mint family have square-sided stems, and leaves arranged in opposite pairs. The flowers often grow in closely packed whorls around the stems, with the lowest flowers opening first.

Leaves in opposite pairs

Whorl of flowers

Square-sided stem

MOTHERWORT

Flower anatomy

Mint family flowers have a wide range of colors, from white to pink, yellow, and blue. They are typically two-lipped. The upper lip often forms a hood, while the lower lip makes up a landing platform for visiting insects. Each flower usually produces four seeds.

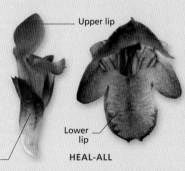

Upper lip

Lower lip

Sepals form funnel-shaped tube

HEAL-ALL

Large Flowers

Plants with large yellow or orange flowers can be found in a variety of habitats—from ponds and slow-moving water, to dry fields and roadsides.

BULLHEAD LILY

Found in ponds and slow-moving water. Its bulbous flowers bloom from May to September, growing just above the water surface. The large leaves, which are notched at the base, float on the water surface.

Cuplike flower

Heart-shaped leaf

↕ Up to 12 in/30 cm above water

MARSH MARIGOLD

Creeping plant of wet meadows, swamps, and streamsides. Its five-petaled flowers bloom from April to June. The kidney-shaped leaves grow on thick, hollow, branching stems.

Saucer-shaped flower

Five overlapping petals

Kidney-shaped leaf

↕ Up to 24 in/60 cm

CALIFORNIA POPPY

Colorful plant of open areas, grassy hillsides, and dry soil along roadsides. In bloom from February to September, its showy, bright flowers have deep orange petals that overlap to form a cup. The blue-green leaves are divided into lobed segments.

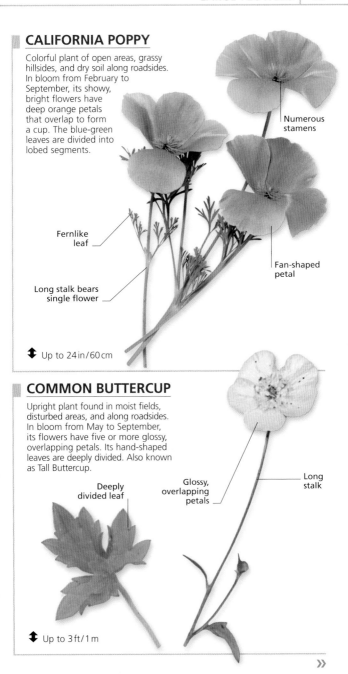

Numerous stamens

Fernlike leaf

Long stalk bears single flower

Fan-shaped petal

✦ Up to 24 in/60 cm

COMMON BUTTERCUP

Upright plant found in moist fields, disturbed areas, and along roadsides. In bloom from May to September, its flowers have five or more glossy, overlapping petals. Its hand-shaped leaves are deeply divided. Also known as Tall Buttercup.

Deeply divided leaf

Glossy, overlapping petals

Long stalk

✦ Up to 3 ft/1 m

»

YELLOW FLAG

Showy plant found in marshy areas and along streamsides. In bloom from June to August. The yellow flowers have three backward-curving sepals and three narrow, upright petals. The pointed, swordlike leaves are often taller than the flower stem.

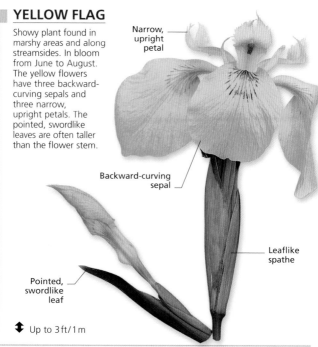

Narrow, upright petal

Backward-curving sepal

Leaflike spathe

Pointed, swordlike leaf

⬍ Up to 3 ft / 1 m

COMMON ST. JOHN'S WORT

Found in fields, waste places, and along roadsides. Its bright yellow flowers bloom from June to September, and have five petals edged with black dots. The small, elliptical leaves have translucent dots.

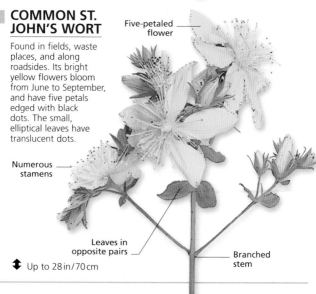

Five-petaled flower

Numerous stamens

Leaves in opposite pairs

Branched stem

⬍ Up to 28 in / 70 cm

SPOTTED TOUCH-ME-NOT

Tall, leafy plant of wet, shady areas; also known as
Jewelweed. The flowers bloom from July to October.
Their orange petals are covered in brown spots and
curve into a mouthlike opening at one end and taper
to a point at the other. The egg-shaped leaves
are toothed. The ripe seedpods
explode when touched.

Toothed
leaf margin

Curved,
mouthlike
opening

Hooked
spur

✦ Up to 5 ft/1.5 m

PRICKLY
PEAR CACTUS

Waxy,
yellow
flower

The only cactus widespread in
eastern North America, this plant
is found in sandy, rocky areas. The
showy yellow flowers bloom
from May to August and
often have a reddish center.
The fleshy, oval leaves,
called pads, form low-
growing, thorny clumps.

Green, scaly
flower bud

Reddish brown
barbed bristle

✦ Up to 12 in/30 cm

»

BUTTER-AND-EGGS TOADFLAX

Widespread plant of roadsides and dry fields. Its pale yellow flowers, similar to a snapdragon, have orange lower lips and a spur at the bottom. In bloom from June to October, they grow in oblong clusters on stems. The numerous leaves are long and narrow.

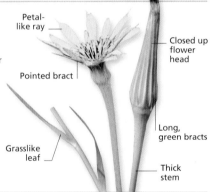

Downward-pointing spur

Long, narrow leaf

Oval seed capsule

 Up to 3 ft / 1 m

YELLOW GOATSBEARD

Found in fields and waste places. Its flower heads are encircled by several long, green bracts at the base. In bloom from May to August, they open early in the day and close by afternoon. The grasslike leaves cling to the stems.

Petal-like ray

Pointed bract

Grasslike leaf

Closed up flower head

Long, green bracts

Thick stem

 Up to 3 ft / 1 m

ORANGE HAWKWEED

Low-growing plant found in fields and on front lawns. Its showy, bright orange flowers have a yellowish center. They bloom from June to September, on a single, hairy stem. The elliptical to lance-shaped leaves form a basal rosette.

Yellowish center

Dense, flat-topped flower head

Hairy stem

 Up to 24 in / 60 cm

COMMON MULLEIN

Statuesque plant found in waste places and along roadsides. Its flowers grow in a densely packed, towering spike. They bloom from June to September. The long leaves are woolly and thick, and form a rosette at the plant's base.

Five-petaled flower

Rounded petal

Densely packed spike

Gray-green stem leaf

⬍ Up to 7 ft/2.1m

PRAIRIE CONEFLOWER

Upright plant of prairies and dry woods. Its flowers have a brown, rounded central cone and long, drooping, yellow petals. They bloom from June to September. The leaves are divided into lance-shaped, toothed segments.

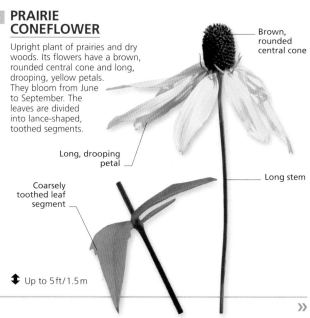

Brown, rounded central cone

Long, drooping petal

Long stem

Coarsely toothed leaf segment

⬍ Up to 5 ft/1.5 m

»

COMMON SUNFLOWER

Found in prairies and along roadsides.
In bloom from July to October, its flower
heads grow on rough, erect stems. They
have a large, dark central disk surrounded
by yellow petal-like rays. The heart-shaped
or spadelike leaves are rough and toothed.

Dark central
disk

Large flower
head

Petal-like
ray

Whorl of
bracts

Rough,
toothed leaf

Rough
stem

⬍ Up to 12 ft / 3.7 m

DANDELION

Found in fields, lawns, and along roadsides. In bloom from March to September, its yellow, disk-shaped flower heads grow on long stems. The flower heads are made up of numerous petal-like rays. The leaves are deeply toothed.

Disk-shaped flower head

Long, hollow stem

Deeply toothed leaf

Petal-like ray

✦ Up to 18 in/45 cm

BLACK-EYED SUSAN

Rough-stemmed plant of prairies, fields, and open woods. Its flower heads have a dark central disk surrounded by many petal-like rays. They grow from June to October. The lance-shaped leaves are rough and hairy.

Daisylike flower head

Dark central disk

Elongated bract

Rough, hairy leaf

Rough stem

✦ Up to 3 ft/1 m

»

JERUSALEM ARTICHOKE

Found in fields and woodland.
Its flower heads, like those of
many sunflowers, have a yellow
central disk surrounded by 10
or more yellow petal-like rays.
They bloom from August
to October. The lance-shaped
leaves are thick and toothed.

Sharp-pointed
bract

Yellow
central disk

Three-
veined leaf

Thick, rough
stem

✦ Up to 10 ft/3 m

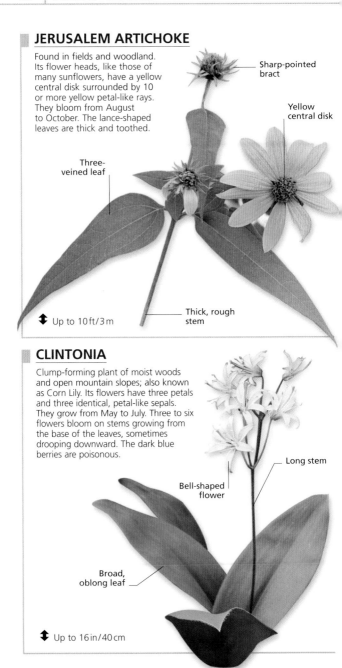

CLINTONIA

Clump-forming plant of moist woods
and open mountain slopes; also known
as Corn Lily. Its flowers have three petals
and three identical, petal-like sepals.
They grow from May to July. Three to six
flowers bloom on stems growing from
the base of the leaves, sometimes
drooping downward. The dark blue
berries are poisonous.

Long stem

Bell-shaped
flower

Broad,
oblong leaf

✦ Up to 16 in/40 cm

PERFOLIATE BELLWORT

Low-growing plant found in moist woodland. In bloom from May to June, its yellowish green, fragrant flowers droop forward. They grow singly on smooth stems and the petals have orange bumps on the inside. The lance-shaped leaves are pierced by the stem at the base.

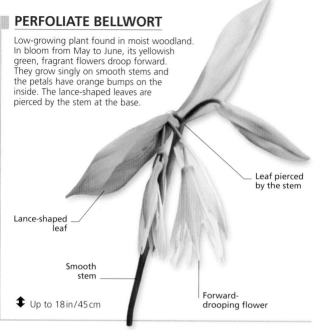

Leaf pierced by the stem

Lance-shaped leaf

Smooth stem

Forward-drooping flower

✦ Up to 18 in/45 cm

LARGE-FLOWERED BELLWORT

Found in moist, shady woodland. In bloom from April to June, its bell-shaped, drooping flowers have slightly twisted petals. Unlike Perfoliate Bellwort, they are smooth on the inside. The egg-shaped or oblong leaves clasp the stems and have a whitish down on the underside.

Smooth stem

Orange–yellow flower

Distinctly veined leaf

✦ Up to 20 in/50 cm

4 RED–PINK FLOWERS

While pure red is a rare color in wildflowers, pink is much more common and widespread. Pink flowers sometimes grade into purple, so look in Chapter 5 (pp.86–103) if you cannot spot a pink flower here.

Small Flowers

Plants with small red or pink flowers sometimes have them growing singly, but in most they are grouped together in clusters or flower heads.

PENNSYLVANIA SMARTWEED

Found in waste places and moist fields. In bloom from May to October, its small, bright pink flowers grow in clusters on sticky, hairy stems. The lance-shaped leaves form a distinctive cylindrical sheath at the base where they join the stems.

Small, bright pink flower

Lance-shaped leaf

Sticky, hairy stem

✛ Up to 4 ft / 1.2 m

SPREADING DOGBANE

Bushy plant found on the edge of dry woodland and along roadsides. Its small, bell-shaped, pink flowers nod forward, growing from where the leaf joins the main stem. They bloom from June to July. The smooth, oval leaves are bluish green.

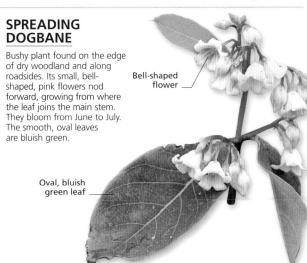

Bell-shaped flower

Oval, bluish green leaf

✛ Up to 4 ft / 1.2 m

COMMON MILKWEED

Tall, downy plant common along roadsides. In bloom from June to August, its pinkish flowers droop slightly. They form globe-shaped flower heads. The oblong leaves produce a toxic, milky substance when bruised.

Oblong leaf

Prominent midrib

Globe-shaped flower head

✦ Up to 6ft/1.8m

SWAMP MILKWEED

Perennial plant found in swamps and wet fields. Its pinkish or ruby-red flowers have a raised crown divided into five parts, and five downward-curving petals. They bloom from June to September. The lance-shaped, opposite leaves taper to a point at the tip.

Rounded flower cluster

Deeply veined leaf

✦ Up to 4ft/1.2m

PEPPERMINT

Widespread plant found near wet meadows, marshes, and along roadsides. Its pinkish purple flowers bloom on short spikes from June to October. The lance-shaped, opposite leaves are sharply toothed.

Pinkish purple flower

Sharply toothed leaf margin

Hairy stem

✦ Up to 3ft/1m

»

SWEET JOE-PYE WEED

Clump-forming plant of moist woodland and prairies. Its fragrant, pinkish to lavender flowers grow in domed clusters, consisting of dozens to hundreds of flower heads. Each flower head has about six tubular flowers. They bloom from July to September. The long leaves grow in whorls of four.

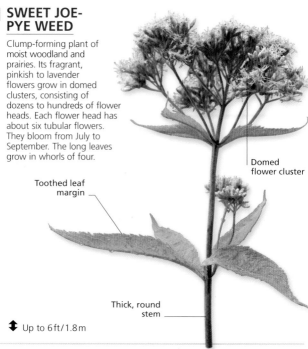

Domed flower cluster

Toothed leaf margin

Thick, round stem

✦ Up to 6 ft / 1.8 m

VALERIAN

Perennial plant found in waste places and along roadsides. Its tiny, pinkish or white flowers grow in rounded clusters from June to July. The leaves are divided into lance-shaped, toothed leaflets, and are slightly hairy on the underside.

Lance-shaped leaflet

Trumpetlike flower

✦ Up to 4 ft / 1.2 m

Medium-sized Flowers

Medium-sized red or pink flowers grow on a variety of plants, including a number of pea family members, such as Crown Vetch.

WILD FOUR O'CLOCK

Found in prairies and waste places. The purple to pink flowers bloom from June to October. They have five notched petals and grow in wide, green, cuplike bracts. The opposite leaves are lance- to egg-shaped, with a heart-shaped or rounded base.

Notched petal

Square-sided stem

Cuplike bract

✚ Up to 4 ft / 1.2 m

COMFREY

Robust, hairy plant found in waste places and along roadsides. Its pink, purple, or creamy white flowers bloom from June to September. They grow in nodding clumps on backward-curling stems, at the base of a pair of oval- to lance-shaped, rough-textured leaves.

Tubular flower

Rough-textured leaf

✚ Up to 3 ft / 1 m

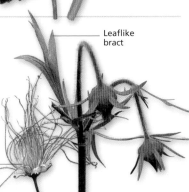

PRAIRIE SMOKE

Hairy plant growing in woodland and prairies. Its reddish brown flowers bloom from April to June. When in fruit, the open flowers resemble a feather duster. The coarse, hairy leaves are divided into several leaflets, with notched tips.

Leaflike bract

Tufted fruit

✚ Up to 16 in / 40 cm

»

RED CLOVER

Low-growing plant of lawns and fields. Its small, fragrant, red or pinkish flowers grow from May to September. They turn brown and papery with age. The leaves are divided into three leaflets at the end of short stems.

Oval leaflet

Dense, rounded flower head

Soft hairs on stem

V-shaped mark on leaflet

✤ Up to 24 in/60 cm

RABBIT-FOOT CLOVER

Low-growing plant of dry, sandy, or gravelly areas. Its pinkish or gray flowers bloom from May to October. The flower heads grow on erect, hairy stems. Its long, narrow, paddle-shaped leaves are toothed at the tips.

Fuzzy, cylindrical flower head

Long, narrow leaf

Toothed tip

Erect, hairy stem

✤ Up to 16 in/40 cm

ALSIKE CLOVER

Found in fields and along roadsides. In bloom from May to October, its showy pea flowers are dull white above and pinkish below. Each flower head grows on a single stalk branching from the main stem. The three-part leaves are unmarked, unlike those of the Red Clover.

Rounded flower head

Toothed, three-part leaf

Long flower stalk

Smooth, thin stem

✦ Up to 24 in/60 cm

CROWN VETCH

Found in fields, waste places, and along roadsides. In bloom from June to August, its pea flowers are pink and white. The leaves are divided into numerous paired, oval leaflets.

Crownlike flower cluster

Slender stem

Paired, oval leaflets

✦ Up to 24 in/60 cm

Mustard Family

There are around 3,500 species in this family worldwide, with many occuring in North America. They have a telltale feature: four petals, arranged in a cross.

Mustard family plants, or crucifers, grow in a variety of habitats, from fields and waysides to rocky coasts and cliffs. They include some common weeds, such as Shepherd's Purse, as well as the wild ancestors of kale, cabbage, broccoli, and other crops.

Forest dweller
Garlic Mustard (above) is one of several mustard family plants found in woodland. It forms dense clumps, and is the dominant species on many forest floors in eastern North America.

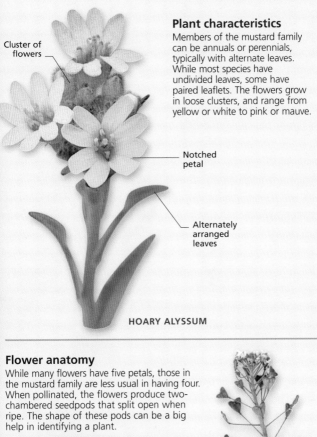

Plant characteristics

Members of the mustard family can be annuals or perennials, typically with alternate leaves. While most species have undivided leaves, some have paired leaflets. The flowers grow in loose clusters, and range from yellow or white to pink or mauve.

Cluster of flowers

Notched petal

Alternately arranged leaves

HOARY ALYSSUM

Flower anatomy

While many flowers have five petals, those in the mustard family are less usual in having four. When pollinated, the flowers produce two-chambered seedpods that split open when ripe. The shape of these pods can be a big help in identifying a plant.

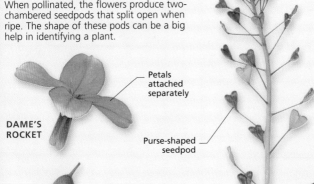

Petals attached separately

DAME'S ROCKET

Purse-shaped seedpod

Oblong seedpod

HOARY ALYSSUM

SHEPHERD'S PURSE

CANADA THISTLE

Invasive plant found in pastures, fields, and along roadsides. Its pale magenta, white, or lavender flower heads are covered with spine-tipped bracts. The flower heads bloom from June to October at the top of smooth stems. The grayish green, lance-shaped leaves are covered with spiny hairs.

Tubular floret

Lance-shaped leaf

Spine-tipped bract

Smooth stem

⬍ Up to 5 ft/1.5 m

SPOTTED CORALROOT

Found in moist, deciduous and coniferous forests, near decaying matter. In bloom from May to August, the flowers have three petal-like sepals above, and two longer ones on the sides. The tonguelike lower white petal has crimson spots. The leaves are tubular sheaths found near the plant's base.

Upper sepal

Longer side sepal

Brownish yellow stalk

⬍ Up to 20 in/50 cm

SPRING BEAUTY

Low-growing plant of moist woodland and thickets. Its pinkish white flowers, with dark pink vertical stripes across the petals, bloom from March to May. The lance-shaped, opposite leaves typically grow in a pair at the midstem.

Lance-shaped leaf

Pink anther

Dark pink, vertical stripes

⬍ Up to 12 in/30 cm

Large Flowers

This group includes some of the most colorful and showy wildflowers—such as Pasture Rose, Indian Paintbrush, and the unmistakable Blanket Flower.

WILD GINGER

Low-growing plant of moist woodland. Its solitary, ruby-red flowers bloom from April to May. They sit at ground level, in the joints of twin leaf stalks. The stem and leaves are covered in fine hairs.

Fuzzy leaf stalk

Ruby-red flower

Heart-shaped leaf

✤ Up to 12 in / 30 cm

TOADSHADE TRILLIUM

Common plant of moist woodland. Its ruby-red or maroon flowers have three closed, erect petals, which sit directly over whorls of three leaves. They bloom from April to June. The leaves have a blotchy pattern of light and dark green.

Stalkless flower

Blotchy pattern on leaf surface

✤ Up to 12 in / 30 cm

FIRE PINK

Upright plant of open woodland thickets and sandy slopes. Its bright red flowers have five heavily cleft petals. In bloom from April to June, they grow in loose groups. The leaves are arranged in pairs on the upper part of the stem, and in tufts at the base.

Five-petaled flower

Lance-shaped leaf

Slender stalk

Tube of sepals

✤ Up to 24 in / 60 cm

»

COLUMBINE

Hardy plant of rocky, wooded areas and open slopes. Its showy red-and-yellow flowers bloom from April to July. They droop forward on delicate stems and have five curved, upward-pointing spurs. The leaves are divided into three-lobed leaflets.

Delicate stem

Upward-pointing spur

Three-lobed leaflet

✢ Up to 24 in/60 cm

DAME'S ROCKET

Stately plant found along roadsides and woodland edges. Its flowers can be purple, pink, or white. In bloom from May to July, they grow in clusters of up to 30. The finely toothed, hairy leaves are wider near the base, tapering to a sharp point at the tip.

Four-petaled flower

Finely toothed leaf

✢ Up to 4 ft/1.2 m

PASTURE ROSE

Found in open woods and rocky pastures. In bloom from June to July, its large, pink flowers have a bright yellow center. The leaves have five to seven toothed leaflets, arranged alternately on slightly hairy stems. Like most roses, it has a thorny main stem. Also known as Carolina Rose.

Five-petaled flower

Toothed leaflet

✢ Up to 3 ft/1 m

FIREWEED

Showy plant growing along roadsides and recently burned areas. In bloom from July to September, its deep pink, four-petaled flowers form tall, loose spikes. The narrow, lance-shaped leaves are slightly wavy at the edges.

Four-petaled flower

Flower bud

Prominent stamens

✦ Up to 7 ft / 2.1 m

WILD BERGAMOT

Showy, aromatic plant found in dry fields and woodland borders. In bloom from June to September, its tubular, purplish pink to lavender flowers grow in rounded clusters on a pincushionlike flower head. The lance-shaped, opposite leaves are slightly toothed.

Tubular flower

Pincushionlike flower head

Lance-shaped leaf

Toothed leaf margin

Square-sided stem

✦ Up to 4 ft / 1.2 m

»

CARDINAL FLOWER

Upright plant found in damp soil, especially near creeks and streams. Its striking bright red flowers bloom from July to September. They have two upper petals and three spreading lower ones attached to a tubular body. The lance-shaped, toothed leaves are arranged alternately on the stems.

Narrow, leaflike bract

Spreading lower petal

Tubular body

↕ Up to 4 ft / 1.2 m

HEDGE BINDWEED

Creeping plant found in damp areas along streams and roadsides. Its trumpetlike flowers are pinkish with white stripes. They bloom from May to September, emerging from pale green bracts. The arrow-shaped or triangular leaves grow along the length of the vine.

Arrow-shaped leaf

Pale green bract

Trumpetlike flower

↔ Stems up to 10 ft / 3 m long

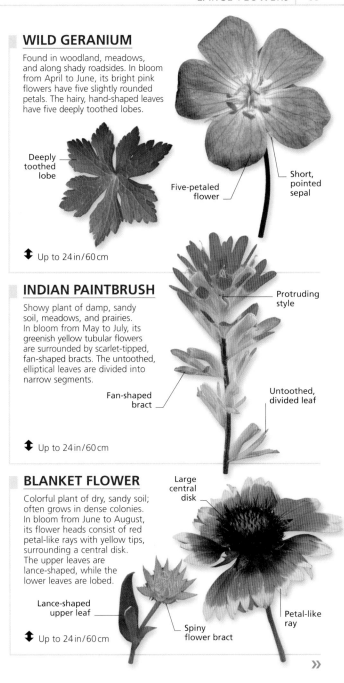

WILD GERANIUM

Found in woodland, meadows, and along shady roadsides. In bloom from April to June, its bright pink flowers have five slightly rounded petals. The hairy, hand-shaped leaves have five deeply toothed lobes.

Deeply toothed lobe

Five-petaled flower

Short, pointed sepal

✦ Up to 24 in/60 cm

INDIAN PAINTBRUSH

Showy plant of damp, sandy soil, meadows, and prairies. In bloom from May to July, its greenish yellow tubular flowers are surrounded by scarlet-tipped, fan-shaped bracts. The untoothed, elliptical leaves are divided into narrow segments.

Protruding style

Fan-shaped bract

Untoothed, divided leaf

✦ Up to 24 in/60 cm

BLANKET FLOWER

Colorful plant of dry, sandy soil; often grows in dense colonies. In bloom from June to August, its flower heads consist of red petal-like rays with yellow tips, surrounding a central disk. The upper leaves are lance-shaped, while the lower leaves are lobed.

Large central disk

Lance-shaped upper leaf

Spiny flower bract

Petal-like ray

✦ Up to 24 in/60 cm

»

PURPLE CONEFLOWER

Showy plant of prairies and dry, open clearings. In bloom from July to October, its flower heads can have up to 20 slightly notched purple-pink, petal-like rays. These are arranged around a spiny, brownish orange central disk. The lance-shaped leaves are broad at the base and slightly folded toward the tip.

Spiny central disk

Notched petal-like ray

Erect, hairy stem

Leaf edge curls inward

Up to 5 ft / 1.5 m

BULL THISTLE

Prickly plant of roadsides, pastures, and waste places. Its rose-purple flowers grow from June to September. They are surrounded by sharp, yellow-tipped bracts. The spear-shaped leaves are long, lobed, and covered with spines.

Dome-shaped flower head

Spiny bract

Spear-shaped leaf

Spiny wing along stem

✦ Up to 6 ft / 1.8 m

SPOTTED KNAPWEED

Invasive plant found in fields, waste places, and along roadsides. Its thistlelike, lavender-pink flowers grow from June to August. The centermost bracts are black-tipped. The upper leaves are small, while those at the base are larger and deeply divided.

Lavender-pink flower head

Ripening seedhead

Prickly, black-tipped bract

Small upper leaf

Wiry stem

✦ Up to 4 ft / 1.2 m

5 PURPLE–BLUE FLOWERS

Some of the most eye-catching plants have purple or blue flowers. They include some well-known spring favorites, such as forget-me-nots and violets, as well as common flowers of late summer, which bring the season to a close.

Small Flowers

Ranging from sky-blue to violet and purple, the flowers in this group are common in grassy areas and along roadsides.

LEAD PLANT

Found in dry soil and prairies. Its flowers grow in dense, spikelike clusters and bloom from May to August. The purple flowers have protruding, orange-tipped stamens. The leaves are divided into 15–45 oval leaflets, which are covered with grayish, woolly fuzz.

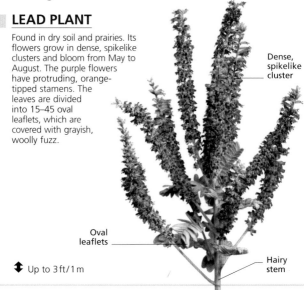

Dense, spikelike cluster

Oval leaflets

Hairy stem

⬍ Up to 3 ft / 1 m

BLUE-EYED GRASS

Petite, stiff, and grasslike plant of moist, open, and grassy areas. Its small, purple flowers have three striped petals and three identical sepals surrounding a yellow center. Each petal ends in a tiny, pointed tip. The flowers bloom from March to July.

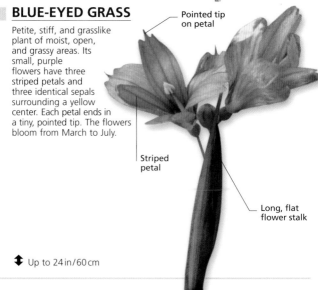

Pointed tip on petal

Striped petal

Long, flat flower stalk

⬍ Up to 24 in / 60 cm

WILD LUPINE

Showy plant found in dry, open woodland and fields. Its flowers bloom from April to July in erect, elongated clusters. They range from white to pink and purple. The hand-shaped, divided leaves radiate out from a central point on the long stems.

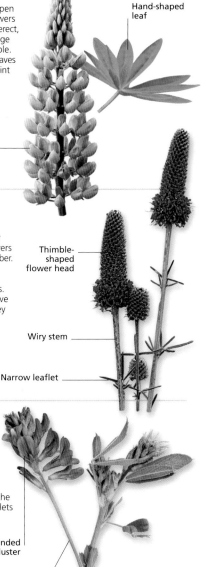

Hand-shaped leaf

Elongated flower cluster

✤ Up to 24 in/60 cm

PURPLE PRAIRIE CLOVER

Found in prairies and on dry hillsides. Its tiny, purple flowers bloom from June to September. They grow upward from the bottom of a spike and form thimble-shaped flower heads. The leaves are divided into five or more narrow leaflets. They are densely packed near the base of the plant and sparse along the stem.

Thimble-shaped flower head

Wiry stem

Narrow leaflet

✤ Up to 3 ft/1 m

ALFALFA

Found along roadsides and in waste places. In bloom from May to October, the purple flowers grow in rounded clusters on long stalks, which shoot up from where the leaves join the stem. The lance-shaped leaflets are grouped in threes.

Rounded flower cluster

Long stalk

✤ Up to 3 ft/1 m

»

FORGET-ME-NOT

Sprawling plant found in wet places and along streamsides. Its tiny, light blue flowers have five petals surrounding a golden center. They appear from May to October on small, curving stems, which uncoil as the flowers bloom. The oblong leaves are hairy and mostly stalkless.

Five-petaled flower

Coiled flowering branch

Branched, fleshy stem

Oblong, hairy leaf

✦ Up to 24 in/60 cm

BLUE VERVAIN

Perennial plant found in damp thickets and along roadsides. Its tubular, violet to blue flowers grow in branched spikes at the top of square-sided stems. They bloom from July to September. The lance-shaped leaves are toothed and rough.

Pencil-like spike

Small, tubular flower

Square-sided stem

✦ Up to 6 ft/1.8 m

Medium-sized Flowers

Plants with medium-sized purple or blue flowers range from low-growing species, such as Common Blue Violet, to climbers and clamberers that use other plants for support.

COW VETCH

Found in fields and along roadsides. The tubular, blue to lavender pea flowers bloom from May to August, and grow on one side of a long spike. The divided leaves have 6–15 pairs of narrow leaflets. Also known as Tufted Vetch.

Curling tendril

One-sided flower cluster

Compact flower buds

Paired leaflets

⬍ Up to 5 ft / 1.5 m

»

FLAX

Generally erect plant of waste places and fields. Its pale blue, five-petaled flowers bloom from June to September on slender stems. The narrow, lance-shaped leaves are stalkless.

Broad petal

Stalkless leaf

Pointed flower bud

Leaves alternately arranged on stem

✤ Up to 20 in/50 cm

COMMON BLUE VIOLET

Found in damp woodland and moist meadows. In bloom from March to June, the flowers can be blue, purple, or white. They grow singly on stalks and have dark purple veins running outward from their whitish center.

Five overlapping petals

Flower grows singly

Heart-shaped leaf, with toothed edge

✤ Up to 8 in/20 cm

PURPLE LOOSESTRIFE

Showy, invasive plant found near ponds and waterways, and along roadsides. Spikes of purple-pink flowers bloom from June to September on erect stems. The long, lance-shaped or straplike leaves are rounded to heart-shaped at the base.

Densely packed spike

Opposite, unstalked leaves

✤ Up to 4 ft/1.2 m

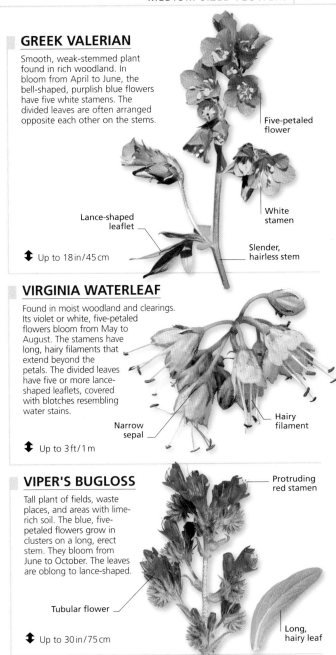

GREEK VALERIAN

Smooth, weak-stemmed plant found in rich woodland. In bloom from April to June, the bell-shaped, purplish blue flowers have five white stamens. The divided leaves are often arranged opposite each other on the stems.

Five-petaled flower

White stamen

Lance-shaped leaflet

Slender, hairless stem

✦ Up to 18 in/45 cm

VIRGINIA WATERLEAF

Found in moist woodland and clearings. Its violet or white, five-petaled flowers bloom from May to August. The stamens have long, hairy filaments that extend beyond the petals. The divided leaves have five or more lance-shaped leaflets, covered with blotches resembling water stains.

Narrow sepal

Hairy filament

✦ Up to 3 ft/1 m

VIPER'S BUGLOSS

Tall plant of fields, waste places, and areas with lime-rich soil. The blue, five-petaled flowers grow in clusters on a long, erect stem. They bloom from June to October. The leaves are oblong to lance-shaped.

Protruding red stamen

Tubular flower

Long, hairy leaf

✦ Up to 30 in/75 cm

»

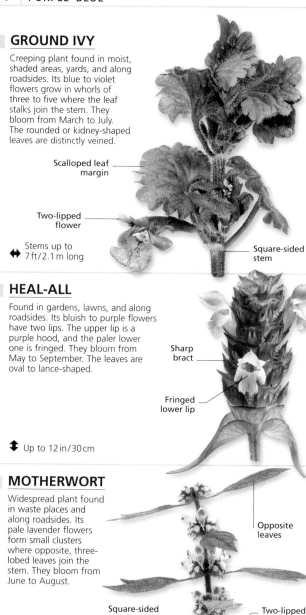

GROUND IVY

Creeping plant found in moist, shaded areas, yards, and along roadsides. Its blue to violet flowers grow in whorls of three to five where the leaf stalks join the stem. They bloom from March to July. The rounded or kidney-shaped leaves are distinctly veined.

Scalloped leaf margin

Two-lipped flower

⬌ Stems up to 7 ft/2.1 m long

Square-sided stem

HEAL-ALL

Found in gardens, lawns, and along roadsides. Its bluish to purple flowers have two lips. The upper lip is a purple hood, and the paler lower one is fringed. They bloom from May to September. The leaves are oval to lance-shaped.

Sharp bract

Fringed lower lip

⬍ Up to 12 in/30 cm

MOTHERWORT

Widespread plant found in waste places and along roadsides. Its pale lavender flowers form small clusters where opposite, three-lobed leaves join the stem. They bloom from June to August.

Opposite leaves

Square-sided stem

Two-lipped flower

⬍ Up to 4 ft/1.2 m

ROUGH BLAZING STAR

Found in prairies and sandy soil. In bloom from August to October, its flower heads consist of star-shaped disk flowers, and form a tufted spike. Tassel-like styles emerge from their center. The narrow, lance-shaped leaves taper to a point at the tip.

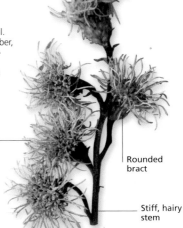

Tassel-like style

Rounded bract

Stiff, hairy stem

⬍ Up to 4 ft / 1.2 m

BITTERSWEET NIGHTSHADE

Climber of thickets and clearings. Its small, star-shaped flowers have deep purple petals surrounding a bright yellow cone of anthers. They bloom from May to September. The lobed leaves are smooth but covered with veins.

Beaklike central cone of anthers

Backward-sweeping petal

Arrow-shaped leaf

⬌ Stems up to 8 ft / 2.4 m long

TEASEL

Tall, prickly plant found in old fields and along roadsides. In bloom from July to October, its tiny, pink to lavender flowers grow in a cylindrical cluster on a thistlelike spike. The lance-shaped leaves are hairy and have a white midrib.

Upward-pointing, spiny bract

Prickly stem

Bristly flower head

Lance-shaped, hairy leaf

⬍ Up to 6 ft / 1.8 m

Pea Family

With nearly 20,000 species across the world, pea family flowers grow in many different habitats. They are easily identifiable because of their unique shape.

Pea family plants, or legumes, are particularly common in grassland, along hedgerows, and around the edges of fields. While there are some woody species (such as Lead Plant) in this family, many are soft-stemmed climbers that use other plants for support.

Purple parade

Wild Lupine (above) was once thought to deplete the soil. Like all pea family plants, however, it actually enriches the ground it grows in. Its typical pea flowers produce hairy seedpods that snap open in warm weather, scattering the seeds far from the parent plant.

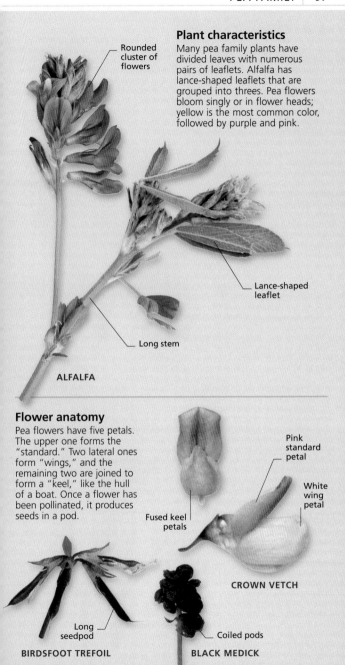

Plant characteristics

Many pea family plants have divided leaves with numerous pairs of leaflets. Alfalfa has lance-shaped leaflets that are grouped into threes. Pea flowers bloom singly or in flower heads; yellow is the most common color, followed by purple and pink.

Rounded cluster of flowers

Lance-shaped leaflet

Long stem

ALFALFA

Flower anatomy

Pea flowers have five petals. The upper one forms the "standard." Two lateral ones form "wings," and the remaining two are joined to form a "keel," like the hull of a boat. Once a flower has been pollinated, it produces seeds in a pod.

Pink standard petal

White wing petal

Fused keel petals

CROWN VETCH

Long seedpod

BIRDSFOOT TREFOIL

Coiled pods

BLACK MEDICK

Large Flowers

Some large purple or blue flowers appear in early spring. Others—such as Bottle Gentian and Chicory—bloom as summer comes to an end.

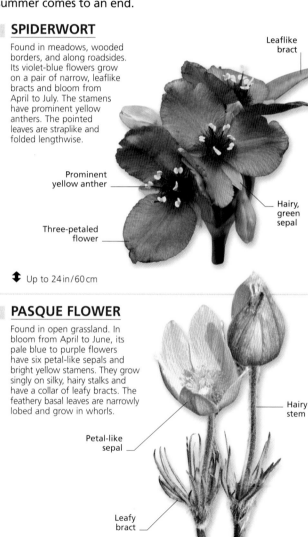

SPIDERWORT

Found in meadows, wooded borders, and along roadsides. Its violet-blue flowers grow on a pair of narrow, leaflike bracts and bloom from April to July. The stamens have prominent yellow anthers. The pointed leaves are straplike and folded lengthwise.

Leaflike bract

Prominent yellow anther

Three-petaled flower

Hairy, green sepal

✦ Up to 24 in/60 cm

PASQUE FLOWER

Found in open grassland. In bloom from April to June, its pale blue to purple flowers have six petal-like sepals and bright yellow stamens. They grow singly on silky, hairy stalks and have a collar of leafy bracts. The feathery basal leaves are narrowly lobed and grow in whorls.

Hairy stem

Petal-like sepal

Leafy bract

✦ Up to 16 in/40 cm

BIRDSFOOT VIOLET

Found in sandy fields and open woodland. Its pale purple to violet-blue flowers bloom from March to June. They have five, slightly irregular petals, the lower of which fades to white. The deeply lobed, fan-shaped leaves are divided into narrow segments.

Irregular petal

Orange anthers

Lower petal fades to white

Deeply lobed, fan-shaped leaf

✦ Up to 10 in / 25 cm

PERIWINKLE

Low-growing, ground-covering plant found along roadsides, in abandoned areas, and in gardens. Its five-petaled, purplish blue flowers have a whitish star in the center. They bloom from April to May. The oval, opposite leaves are dark green and waxy. Also known as Myrtle.

Whitish star in center

Waxy leaf surface

Oval leaf

✦ Up to 8 in / 20 cm

WILD BLUE PHLOX

Found in rich woodland and fields. Its pale blue flowers bloom from April to June. The petals are fused together at the base, forming a tube at the flower's center. The flowers form loose, slightly rounded clusters on a single stem. The oval to lance-shaped leaves are unstalked.

Five-petaled flower

Single, sticky stem

Rough, hairy leaf

✦ Up to 20 in / 50 cm

»

VIRGINIA BLUEBELLS

Upright plant of floodplains and
moist woodland; also known as
Mertensia. Its pink buds grow into
nodding, trumpetlike, blue flowers.
They bloom from March to June.
The oval, untoothed leaves are
larger at the base.

Oval,
untoothed leaf

Trumpetlike
flower

Long,
slender stem

✦ Up to 24 in/60 cm

CHICORY

Found in fields, waste places, and
along roadsides. Its purplish blue
flower heads have square-tipped,
fringed petal-like rays. They bloom
from June to October. The oblong
to lance-shaped basal leaves grasp
the stems.

Stiff stem

Square-tipped
petal-like ray

Lance-shaped
basal leaf

✦ Up to 4 ft/1.2 m

BOTTLE GENTIAN

Perennial plant found in moist
woodland and meadows. In
bloom from August to October,
its purple flowers have five
fused petals with pleats
between each. The petals
never open and resemble large
buds. The slightly oval, smooth
leaves taper to a point at the tip.

Fused
petals

Leaf tapers
to a point

Slender
stem

✦ Up to 24 in/60 cm

EUROPEAN BELLFLOWER

Clump-forming plant found in shady deciduous woodland and along roadsides. Its bell-shaped, blue to purple flowers bloom from June to October. They have five petals and grow on one side of a long stalk. The lance- or heart-shaped leaves are deeply toothed.

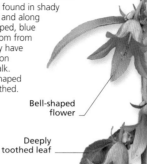

Bell-shaped flower

Deeply toothed leaf

✦ Up to 3 ft / 1 m

HAREBELL

Found on rocky slopes, in meadows, and along shorelines. Its bell-shaped, violet flowers grow from June to October. They have five pointed petals that flare slightly. The grasslike leaves are long and narrow.

Bell-shaped flower

Long, narrow leaf

Threadlike stem

✦ Up to 20 in / 50 cm

CLUSTERED BELLFLOWER

Upright, spreading plant found in open woodland and grassy areas along roadsides. Its deep purple flowers grow in clusters on erect stems. They bloom from June to September. The oval to lance-shaped leaves are rounded at the base.

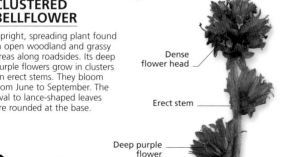

Dense flower head

Erect stem

Deep purple flower

✦ Up to 24 in / 60 cm

»

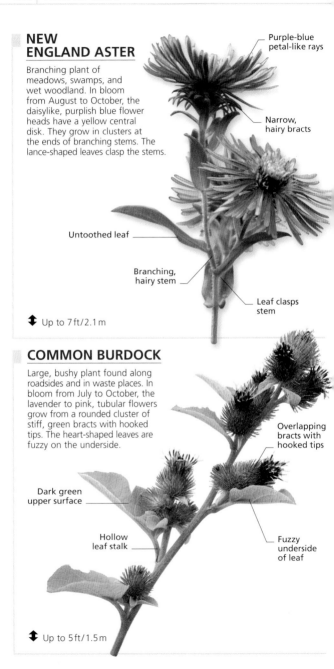

NEW ENGLAND ASTER

Branching plant of meadows, swamps, and wet woodland. In bloom from August to October, the daisylike, purplish blue flower heads have a yellow central disk. They grow in clusters at the ends of branching stems. The lance-shaped leaves clasp the stems.

Purple-blue petal-like rays

Narrow, hairy bracts

Untoothed leaf

Branching, hairy stem

Leaf clasps stem

↕ Up to 7 ft / 2.1 m

COMMON BURDOCK

Large, bushy plant found along roadsides and in waste places. In bloom from July to October, the lavender to pink, tubular flowers grow from a rounded cluster of stiff, green bracts with hooked tips. The heart-shaped leaves are fuzzy on the underside.

Overlapping bracts with hooked tips

Dark green upper surface

Hollow leaf stalk

Fuzzy underside of leaf

↕ Up to 5 ft / 1.5 m

PURPLE FRINGED ORCHID

Showy plant of wet meadows and moist woodland. Its lavender-pink flowers bloom from June to August. The flowers have two erect upper petals and three fan-shaped lower petals. They also have three petal-like sepals—an erect upper one and two oval side sepals. The leaves are egg- to lance-shaped. The upper leaves are small; lower ones are longer and sheath the stem.

Elongated cluster of flowers

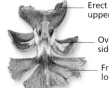

Erect upper petal

Oval side sepal

Fringed lower petal

Backward-pointing spur

✦ Up to 4 ft / 1.2 m

BLUE FLAG

Found in swamps, marshes, and along shorelines. Its flowers bloom from May to August. They have three sepals that are veined and yellow near the base, and three narrower, erect petals. The lance-shaped leaves are very long and thick.

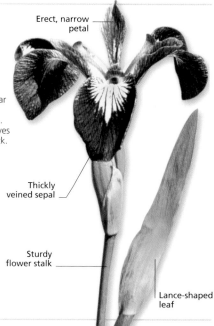

Erect, narrow petal

Thickly veined sepal

Sturdy flower stalk

Lance-shaped leaf

✦ Up to 3 ft / 1 m

FLOWER GALLERY

This gallery shows the wildflowers already profiled in the book, grouped by family. A family consists of closely related genera, which are, in turn, made up of related species that often look similar. You can use this gallery if you think you know the family the flower is in, or to find out which one it belongs to. Then go to its profile page to learn all about it.

ALISMATACEAE
Water Plantain Family

Broad-leaved
Arrowhead
p.40

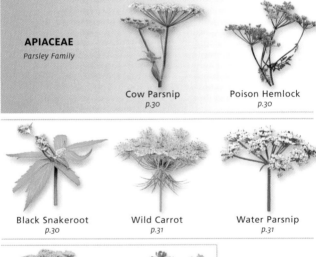

APIACEAE
Parsley Family

Cow Parsnip
p.30

Poison Hemlock
p.30

Black Snakeroot
p.30

Wild Carrot
p.31

Water Parsnip
p.31

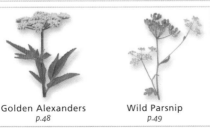

Golden Alexanders
p.48

Wild Parsnip
p.49

APOCYNACEAE
Dogbane Family

Spreading Dogbane
p.70

Periwinkle
p.99

ARACEAE
Arum Family

Jack-in-the-pulpit
p.22

Skunk Cabbage
p.23

ARISTOLOCHIACEAE
Birthwort Family

Wild Ginger
p.79

ASCLEPIADACEAE
Milkweed Family

Butterfly Weed
p.50

Common Milkweed
p.71

Swamp Milkweed
p.71

ASPARAGACEAE
Asparagus Family

Solomon's Seal
p.21

»

»

Yucca
p.45

ASTERACEAE
*Sunflower
Family*

Common Ragweed
p.19

Yarrow
p.32

Pearly Everlasting
p.32

Boneset
p.32

Feverfew
p.37

Daisy Fleabane
p.37

White Lettuce
p.40

Flat-top Aster
p.40

Mayweed
p.42

Oxeye Daisy
p.43

Pineapple Weed
p.50

Lance-leaved Goldenrod
p.51

Canada Goldenrod
p.51

Wild Lettuce
p.51

Common Tansy
p.54

Golden Ragwort
p.55

King Devil
p.55

Yellow Goatsbeard
p.62

Orange Hawkweed
p.62

Prairie Coneflower
p.63

Common Sunflower
p.64

Dandelion
p.65

Black-eyed Susan
p.65

Jerusalem Artichoke
p.66

Sweet Joe-Pye Weed
p.72

Canada Thistle
p.78

Blanket Flower
p.83

Purple Coneflower
p.84

Bull Thistle
p.85

Spotted
Knapweed
p.85

Rough
Blazing Star
p.95

»

Chicory
p.100

New England Aster
p.102

Common Burdock
p.102

BALSAMINACEAE

Touch-me-not Family

Spotted Touch-me-not
p.61

BERBERIDACEAE

Barberry Family

Mayapple
p. 43

BORAGINACEAE

Borage Family

Comfrey
p.73

Forget-me-not
p.90

Viper's Bugloss
p.93

Virginia Bluebells
p.100

BRASSICACEAE

Mustard Family

Hoary Alyssum
p.28

Garlic Mustard
p.29

Shepherd's Purse
p.29

Cutleaf Toothwort
p.36

Wintercress
p.48

Dame's Rocket
p.80

CACTACEAE
Cactus Family

Prickly Pear Cactus
p.61

CAMPANULACEAE
Bellflower Family

Cardinal Flower
p.82

European Bellflower
p.101

Harebell
p.101

Clustered Bellflower
p.101

CARYOPHYLLACEAE
Carnation Family

Common Chickweed
p.28

Bladder Campion
p.34

»

»

Fire Pink
p.79

COLCHICACEAE
Colchicum Family

Perfoliate Bellwort
p.67

Large-flowered Bellwort
p.67

COMMELINACEAE
Spiderwort Family

Spiderwort
p.98

CONVOLVULACEAE
Morning Glory Family

Hedge Bindweed
p.82

CORNACEAE
Dogwood Family

Bunchberry
p.42

CUCURBITACEAE
Cucumber Family

Wild Cucumber
p.33

DIPSACACEAE
Teasel Family

Teasel
p.95

EUPHORBIACEAE
Spurge Family

Leafy Spurge
p.20

FABACEAE
Pea Family

White Sweet Clover
p.29

White Clover
p.36

Black Medick
p.48

Yellow Sweet
Clover
p.49

Birdsfoot Trefoil
p.52

Hop Clover
p.53

Red Clover
p.74

Rabbit-foot
Clover
p.74

Alsike Clover
p.75

Crown Vetch
p.75

Lead Plant
p.88

Wild Lupine
p.89

Purple Prairie Clover
p.89

Alfalfa
p.89

Cow Vetch
p.91

FUMARIACEAE
Fumitory Family

Dutchman's Breeches
p.35

GENTIANACEAE
Gentian Family

Bottle Gentian
p.100

GERANIACEAE
Geranium Family

Wild Geranium
p.83

HYDROPHYLLACEAE
Waterleaf Family

Virginia Waterleaf
p.93

HYPERICACEAE
St. John's Wort Family

Common St. John's Wort
p.60

IRIDACEAE
Iris Family

Yellow Flag
p.60

Blue-eyed Grass
p.88

Blue Flag
p.103

LAMIACEAE
Mint Family

Catnip
p.36

Peppermint
p.71

Wild Bergamot
p.81

Ground Ivy
p.94

Heal-all
p.94

Motherwort
p.94

LILIACEAE
Lily Family

Large-flowered Trillium
p.44

Nodding Trillium
p.44

Clintonia
p.66

Toadshade Trillium
p.79

LINACEAE
Flax Family

Flax
p.92

LYTHRACEAE
Loosestrife Family

Purple Loosestrife
p.92

MONOTROPACEAE
Indian Pipe Family

Indian Pipe
p.35

NYCTAGINACEAE
Four O'Clock Family

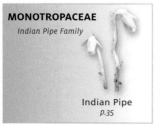

Wild Four O'Clock
p.73

NYMPHAEACEAE
Water Lily Family

Bullhead Lily
p.58

ONAGRACEAE
Evening Primrose Family

Fireweed
p.81

ORCHIDACEAE
Orchid Family

Spotted Coralroot
p.78

Purple Fringed Orchid
p.103

OXALIDACEAE
Wood Sorrel Family

Yellow Wood Sorrel
p.53

PAPAVERACEAE
Poppy Family

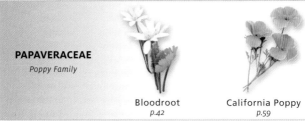

Bloodroot
p.42

California Poppy
p.59

PLANTAGINACEAE
Plantain Family

Common Plantain
p.33

POLEMONIACEAE
Phlox Family

Greek Valerian
p.93

Wild Blue Phlox
p.99

POLYGONACEAE
Buckwheat Family

Curly Dock
p.18

Pennsylvania Smartweed
p.70

PORTULACACEAE
Purslane Family

Spring Beauty
p.78

PRIMULACEAE
Primrose Family

Lance-leaved Loosestrife
p.52

RANUNCULACEAE
Buttercup Family

Early Meadow-rue
p.19

Sharp-lobed Hepatica
p.34

Black Cohosh
p.34

Wood Anemone
p.41

Canada Anemone
p.41

»

»

Marsh Marigold
p.58

Common Buttercup
p.59

Columbine
p.80

Pasque Flower
p.98

ROSACEAE
Rose Family

Common Strawberry
p.35

Rough-fruited
Cinquefoil
p.52

Prairie Smoke
p.73

Pasture Rose
p.80

SCROPHULARIACEAE
Figwort Family

Culver's Root
p.31

Butter-and-eggs
Toadflax
p.62

Common Mullein
p.63

Indian Paintbrush
p.83

SOLANACEAE
Nightshade Family

Bittersweet Nightshade
p.95

URTICACEAE
Nettle Family

Stinging Nettle
p.18

VALERIANACEAE
Valerian Family

Valerian
p.72

VERBENACEAE
Verbena Family

Blue Vervain
p.90

VIOLACEAE
Violet Family

Downy Yellow Violet
p.53

Common Blue Violet
p.92

Birdsfoot Violet
p.99

Scientific Names

Every living species has a scientific name of two Latin words. The first word is the genus, shared by closely related species that often look similar. The second is the specific name. Each two-word combination is unique to the individual species.

Common name	Scientific name	Page
Stinging Nettle	Urtica dioica	18
Curly Dock	Rumex crispus	18
Early Meadow-rue	Thalictrum dioicum	19
Common Ragweed	Ambrosia artemisiifolia	19
Leafy Spurge	Euphorbia esula	20
Solomon's Seal	Polygonatum biflorum	21
Jack-in-the-pulpit	Arisaema atrorubens	22
Skunk Cabbage	Symplocarpus foetidus	23
Hoary Alyssum	Berteroa incana	28
Common Chickweed	Stellaria media	28
Garlic Mustard	Alliaria petiolata	29
Shepherd's Purse	Capsella bursa-pastoris	29
White Sweet Clover	Melilotus alba	29
Cow Parsnip	Heracleum maximum	30
Poison Hemlock	Conium maculatum	30
Black Snakeroot	Sanicula marilandica	30
Wild Carrot	Daucus carota	31
Water Parsnip	Sium suave	31
Culver's Root	Veronicastrum virginicum	31
Yarrow	Achillea millefolium	32
Pearly Everlasting	Anaphalis margaritacea	32
Boneset	Eupatorium perfoliatum	32
Common Plantain	Plantago major	33
Wild Cucumber	Echinocystis lobata	33
Sharp-lobed Hepatica	Hepatica acutiloba	34
Bladder Campion	Silene cucubalus	34
Black Cohosh	Actaea racemosa	34
Dutchman's Breeches	Dicentra cucullaria	35
Indian Pipe	Monotropa uniflora	35
Common Strawberry	Fragaria virginiana	35
White Clover	Trifolium repens	36
Catnip	Nepeta cataria	36
Cutleaf Toothwort	Cardamine concatenata	36
Feverfew	Chrysanthemum parthenium	37
Daisy Fleabane	Erigeron annuus	37
White Lettuce	Prenanthes alba	40
Flat-top Aster	Aster umbellatus	40
Broad-leaved Arrowhead	Sagittaria latifolia	40

Wood Anemone	*Anemone quinquefolia*	41
Canada Anemone	*Anemone canadensis*	41
Mayweed	*Anthemis cotula*	42
Bloodroot	*Sanguinaria canadensis*	42
Bunchberry	*Cornus canadensis*	42
Oxeye Daisy	*Chrysanthemum leucanthemum*	43
Mayapple	*Podophyllum peltatum*	43
Large-flowered Trillium	*Trillium grandiflorum*	44
Nodding Trillium	*Trillium cernuum*	44
Yucca	*Yucca filamentosa*	45
Wintercress	*Barbarea vulgaris*	48
Black Medick	*Medicago lupulina*	48
Golden Alexanders	*Zizia aurea*	48
Yellow Sweet Clover	*Melilotus officinalis*	49
Wild Parsnip	*Pastinaca sativa*	49
Butterfly Weed	*Asclepias tuberosa*	50
Pineapple Weed	*Matricaria matricarioides*	50
Lance-leaved Goldenrod	*Euthamia graminifolia*	51
Canada Goldenrod	*Solidago canadensis*	51
Wild Lettuce	*Lactuca canadensis*	51
Rough-fruited Cinquefoil	*Potentilla recta*	52
Lance-leaved Loosestrife	*Lysimachia lanceolata*	52
Birdsfoot Trefoil	*Lotus corniculatus*	52
Hop Clover	*Trifolium agrarium*	53
Yellow Wood Sorrel	*Oxalis stricta*	53
Downy Yellow Violet	*Viola pubescens*	53
Common Tansy	*Tanacetum vulgare*	54
Golden Ragwort	*Senecio aureus*	55
King Devil	*Hieracium caespitosum*	55
Bullhead Lily	*Nuphar variegatum*	58
Marsh Marigold	*Caltha palustris*	58
California Poppy	*Eschscholzia californica*	59
Common Buttercup	*Ranunculus acris*	59
Yellow Flag	*Iris pseudacorus*	60
Common St. John's Wort	*Hypericum perforatum*	60
Spotted Touch-me-not	*Impatiens capensis*	61
Prickly Pear Cactus	*Opuntia humifusa*	61
Butter-and-eggs Toadflax	*Linaria vulgaris*	62
Yellow Goatsbeard	*Tragopogon pratensis*	62
Orange Hawkweed	*Hieracium aurantiacum*	62
Common Mullein	*Verbascum thapsus*	63
Prairie Coneflower	*Ratibida pinnata*	63
Common Sunflower	*Helianthus annuus*	64
Dandelion	*Taraxacum officinale*	65
Black-eyed Susan	*Rudbeckia hirta*	65
Jerusalem Artichoke	*Helianthus tuberosus*	66
Clintonia	*Clintonia borealis*	66

Perfoliate Bellwort	*Uvularia perfoliata*	67
Large-flowered Bellwort	*Uvularia grandiflora*	67
Pennsylvania Smartweed	*Polygonum pensylvanicum*	70
Spreading Dogbane	*Apocynum androsaemifolium*	70
Common Milkweed	*Asclepias syriaca*	71
Swamp Milkweed	*Asclepias incarnata*	71
Peppermint	*Mentha piperita*	71
Sweet Joe-Pye Weed	*Eupatorium purpureum*	72
Valerian	*Valeriana officinalis*	72
Wild Four O'Clock	*Mirabilis nyctaginea*	73
Comfrey	*Symphytum officinale*	73
Prairie Smoke	*Geum triflorum*	73
Red Clover	*Trifolium pratense*	74
Rabbit-foot Clover	*Trifolium arvense*	74
Alsike Clover	*Trifolium hybridum*	75
Crown Vetch	*Coronilla varia*	75
Canada Thistle	*Cirsium arvense*	78
Spotted Coralroot	*Corallorhiza maculata*	78
Spring Beauty	*Claytonia virginica*	78
Wild Ginger	*Asarum canadense*	79
Toadshade Trillium	*Trillium sessile*	79
Fire Pink	*Silene virginica*	79
Columbine	*Aquilegia canadensis*	80
Dame's Rocket	*Hesperis matronalis*	80
Pasture Rose	*Rosa carolina*	80
Fireweed	*Epilobium angustifolium*	81
Wild Bergamot	*Monarda fistulosa*	81
Cardinal Flower	*Lobelia cardinalis*	82
Hedge Bindweed	*Calystegia sepium*	82
Wild Geranium	*Geranium maculatum*	83
Indian Paintbrush	*Castillleja coccinea*	83
Blanket Flower	*Gaillardia pulchella*	83
Purple Coneflower	*Echinacea purpurea*	84
Bull Thistle	*Cirsium vulgare*	85
Spotted Knapweed	*Centaurea maculosa*	85
Lead Plant	*Amorpha canescens*	88
Blue-eyed Grass	*Sisyrinchium angustifolium*	88
Wild Lupine	*Lupinus perennis*	89
Purple Prairie Clover	*Dalea purpurea*	89
Alfalfa	*Medicago sativa*	89
Forget-me-not	*Myosotis scorpioides*	90
Blue Vervain	*Verbena hastata*	90
Cow Vetch	*Vicia cracca*	91
Flax	*Linum usitatissimum*	92
Common Blue Violet	*Viola papilionacea*	92
Purple Loosestrife	*Lythrum salicaria*	92
Greek Valerian	*Polemonium reptans*	93

Virginia Waterleaf	*Hydrophyllum virginianum*	93
Viper's Bugloss	*Echium vulgare*	93
Ground Ivy	*Glechoma hederacea*	94
Heal-all	*Prunella vulgaris*	94
Motherwort	*Leonurus cardiaca*	94
Rough Blazing Star	*Liatris aspera*	95
Bittersweet Nightshade	*Solanum dulcamara*	95
Teasel	*Dipsacus fullonum*	95
Spiderwort	*Tradescantia virginiana*	98
Pasque Flower	*Pulsatilla patens*	98
Birdsfoot Violet	*Viola pedata*	99
Periwinkle	*Vinca minor*	99
Wild Blue Phlox	*Phlox divaricata*	99
Virginia Bluebells	*Mertensia virginica*	100
Chicory	*Cichorium intybus*	100
Bottle Gentian	*Gentiana andrewsii*	100
European Bellflower	*Campanula rapunculoides*	101
Harebell	*Campanula rotundifolia*	101
Clustered Bellflower	*Campanula glomerata*	101
New England Aster	*Aster novae-angliae*	102
Common Burdock	*Arctium minus*	102
Purple Fringed Orchid	*Habenaria fimbriata*	103
Blue Flag	*Iris versicolor*	103

Glossary

The terms defined here can help you understand wildflowers better and describe them with a greater degree of precision.

Alternate An arrangement where leaves grow singly at different levels along the stem.

Annual A plant that completes its life cycle in less than a year.

Anther The pollen-bearing part at the end of the stamen of a flower.

Axil The angle between two structures, such as the leaf and stem.

Basal Located at the base of the plant.

Bearded Describes a petal that has a tuft or ring of hairs.

Biennial A plant that does not flower or fruit until the second year of its two-year life cycle.

Bract A leaflike organ at the base of a flower stalk.

Carpel The female reproductive organ, consisting of an ovary, a stigma, and a style.

Cultivated A plant specially bred or improved, or grown for its produce.

Disk floret One of the small, tubular flowers at the center of the flower head of members of the sunflower family.

Divided Describes a leaf that is split into distinct leaflets or lobes.

Elliptical Describes a leaf with a wide middle and ends that taper to points.

Family A scientific grouping of closely related plants.

Floret One of a group of small, individual flowers usually clustered together to form a flower head.

Flower head A cluster of florets or larger flowers.

Fruit The ripened dry or fleshy seed-bearing structure of a flowering plant.

Genus (pl. Genera) A category in classification consisting of a group of closely related species, and denoted by the first part of the scientific name, e.g. *Melilotus* in *Melilotus alba*.

Keel petal The lower, fused petals of a pea flower.

Leaflet One of the leaflike structures that make up a divided leaf.

Lip A protruding petal, as found in members of the orchid and mint families.

Lobe An often-rounded part of a divided leaf, formed by incisions toward the midrib.

Midrib The often-prominent central vein on a leaf or leaflike structure.

Nectar A sugary fluid produced by some flowers, which draws insect pollinators.

Node The point from which leaves sprout on a stem.

Opposite Describes leaves growing in pairs along the stem.

Pea flower A flower, usually from the pea family, with sepals fused into a short tube and a distinctive petal arrangement.

Perennial A plant with a life cycle that spans more than two years.

Ray/Ray floret One of the small, flattened flowers that are usually found around the edge of the flower head in members of the sunflower family.

Rhizome A horizontal fleshy stem, usually growing underground.

Rosette A circular cluster of leaves arranged at or near the base of a stem.

Runner A stem that creeps along the ground, forming roots at intervals and, eventually, separate plants.

Sap The watery fluid that carries nutrients and other substances around inside a plant.

Seedhead A flower head in seed.

Sepal One of the separate, and usually green, parts of the flower that enclose the petals.

Spathe A large, hooded bract.

Spur A hollow, cylindrical or pouched structure projecting from a flower, usually containing nectar.

Stamen The male reproductive organ of a flower, consisting of a stalk bearing an anther.

Standard petal The upright, upper petal of a pea flower, often larger than the others.

Stigma The part of the flower that receives pollen.

Stipule A leaflike organ at the base of a leaf stalk.

Style The part of the female reproductive organ that joins the ovary to the pollen-receiving stigma.

Taproot A large, single root that grows vertically downward, and from which other roots sprout.

Tendril A slender organ used by climbing plants to cling to a supporting object.

Toothed Describes a leaf margin with indentations.

Umbel A flat-topped or domed cluster with stalks that rise from a common point.

Weed An invasive plant that grows where it is not wanted—often disrupting the habitat and threatening the survival of existing plants, or affecting crops on farmland.

Whorl An arrangement where leaves grow in rings.

Wing petal One of the lateral petals of many flowers, particularly orchids and pea flowers.

Index

Page numbers in **bold**
indicate main entry.

A

Alexanders, Golden
48, 104
Alfalfa **89**, 97, 111
Alsike Clover **75**,
111
Alyssum, Hoary **28**,
77, 108
anatomy
  flower anatomy 9,
  10
  leaf anatomy 12
  mint family 57
  mustard family 77
  parsley family 25
  pea family 97
  plant anatomy 8
  sunflower family 39
Anemone
  Canada **41**, 115
  Wood **41**, 115
Arrowhead,
  Broad-leaved **40**,
  104
Artichoke, Jerusalem
  66, 107
Aster
  Flat-top **40**, 106
  New England 39,
  102, 108

B

Bellflower
  Clustered **101**, 109
  European **101**, 109
Bellwort
  Large-flowered
  67, 110
  Perfoliate **67**, 110
Bergamot, Wild 56,
  81, 113
Bindweed, Hedge **82**,
  110

Birdsfoot Trefoil 11,
  14, **52**, 97, 111
Birdsfoot Violet **99**,
  117
Bittersweet Nightshade
  9, **95**, 117
Black Cohosh **34**, 115
Black-eyed Susan
  65, 107
Black Medick 14, **48**,
  97, 111
Black Snakeroot **30**,
  104
Bladder Campion **34**,
  109
Blanket Flower 79, **83**,
  107
Blazing Star, Rough 95
Bloodroot 9, 26,
  42, 114
Bluebells, Virginia 11,
  100, 108
Blue-eyed Grass **88**,
  112
Blue Flag **103**, 112
Blue Vervain **90**, 117
Boneset **32**, 106
Bottle Gentian 98,
  100, 112
Broad-leaved
  Arrowhead **40**,
  104
Bugloss, Viper's **93**,
  108
Bull Thistle 10, 39,
  85, 107
Bullhead Lily 9,
  58, 114
Bunchberry 14,
  42, 110
Burdock, Common
  102, 108
Butter-and-eggs
  Toadflax **62**,
  116
buttercup family
  115–16
Buttercup
  Common **59**, 116
  Tall 59
Butterfly Weed **50**,
  105

C

Cabbage, Skunk 22,
  23, 105
Cactus, Prickly Pear **61**,
  109
California Poppy 11,
  14, **59**, 114
Campion, Bladder **34**,
  109
Canada Anemone **41**,
  115
Canada Thistle **78**, 107
Canada Goldenrod **51**,
  106
Cardinal Flower **82**,
  109
Carolina Rose 80
Carrot, Wild 11, 24,
  31, 104
Catnip **36**, 112
Chickweed, Common
  28, 109
Chicory 98, **100**, 108
Cinquefoil,
  Rough-fruited
  52, 116
Clintonia **66**, 113
Clover
  Alsike **75**, 111
  Hop **53**, 111
  Rabbit-foot **74**, 111
  Red **74**, 75, 111
  White **36**, 111
Clustered Bellflower
  101, 109
Cohosh, Black **34**, 115
Columbine 14,
  80, 116
Comfrey **73**, 108
Common Blue Violet
  91, **92**, 117
Common Burdock
  102, 108
Common Buttercup
  59, 116
Common Chickweed
  28, 109
Common Milkweed
  71, 105
Common Mullein 11,
  63, 116

Common Plantain 12, **33**, 114
Common Ragweed **19**, 106
Common St. John's Wort **60**, 112
Common Strawberry 9, **35**, 116
Common Sunflower **64**, 107
Common Tansy **54**, 107
Coneflower
  Prairie **63**, 107
  Purple 10, **84**, 107
Coralroot, Spotted 11, **78**, 114
Corn Lily 66
Corpse Plant 35
Cow Parsnip **30**, 104
Cow Vetch 11, **91**, 111
Crown Vetch 73, **75**, 97, 111
Cucumber, Wild 14, **33**, 110
Culver's Root **31**, 34, 116
Curly Dock 16, **18**, 115
Cutleaf Toothwort **36**, 77, 109

D

Daisy, Oxeye 37, 39, **43**, 106
Daisy Fleabane **37**, 106
Dame's Rocket 77, **80**, 109
Dandelion 11, 14, 38, 39, **65**, 107
Dock, Curly 16, **18**, 115
Dogbane, Spreading **70**, 105
Downy Yellow Violet **53**, 117
Dutchman's Breeches **35**, 112

E

Early Meadow-rue **19**, 115
European Bellflower **101**, 109

F

Feverfew **37**, 106
Fire Pink 11, **79**, 110
Fireweed 9, **81**, 114
Flag
  Blue **103**, 112
  Yellow 14, 46, **60**, 112
Flat-top Aster **40**, 106
Flax **92**, 113
Fleabane, Daisy **37**, 106
Forget-me-not 86, **90**, 108
Four O'Clock, Wild **73**, 113
fruit 14

G

Garlic Mustard **29**, 76, 108
Gentian, Bottle 98, **100**, 112
Geranium, Wild 10, **83**, 112
Ginger, Wild **79**, 105
Goatsbeard, Yellow **62**, 107
Golden Alexanders **48**, 104
Golden Ragwort **55**, 107
Goldenrod
  Canada **51**, 106
  Lance-leaved **51**, 106
Greek Valerian **93**, 115
Ground Ivy **94**, 113
growth, plants 9

H

habitats 14
Harebell **101**, 109
Hawkweed, Orange **62**, 107
Heal-all 57, **94**, 113
Hedge Bindweed **82**, 110
Hemlock, Poison **30**, 104
Hepatica,
  Sharp-lobed 26, **34**, 115
Hoary Alyssum **28**, 77, 108
Hop Clover **53**, 111

I

Indian Paintbrush 79, **83**, 116
Indian Pipe **35**, 113
iris family 112
Ivy, Ground **94**, 113

J

Jack-in-the-pulpit **22**, 105
Jerusalem Artichoke **66**, 107
Jewelweed **61**
Joe-Pye Weed, Sweet **72**, 107

K

King Devil **55**, 107
Knapweed, Spotted **85**, 107

L

Lance-leaved Goldenrod **51**, 106
Lance-leaved Loosestrife **52**, 115

Large-flowered
      Bellwort **67**, 110
Large-flowered Trillium
      44, 113
Lead Plant **88**,
      96, 111
leaf anatomy 12
Leafy Spurge 16, **20**,
      110
leaves 8, **12–13**
Lettuce
   White **40**, 106
   Wild **51**, 106
Lily
   Bullhead 9,
      58, 114
   Corn 66
Loosestrife
   Lance-leaved **52**, 115
   Purple **92**, 113
Lupine, Wild **89**,
      96, 111

M

Marsh Marigold 14,
      58, 116
Marigold, Marsh 14,
      58, 116
Mayapple 14,
      43, 108
Mayweed **42**, 106
Meadow-rue, Early **19**,
      115
Medick, Black 14,
      48, 97, 111
Mertensia 100
Milkweed
   Common **71**, 105
   Swamp **71**, 105
mint family **56–7**,
      112–13
Motherwort 57, **94**,
      113
Mullein, Common 11,
      63, 116
Mustard, Garlic **29**, 76,
      108
mustard family **76–7**,
      108–9
Myrtle 99

N

Nettle, Stinging 9, **18**,
      117
New England Aster 39,
      102, 108
Nightshade,
      Bittersweet 9, **95**,
      117
Nodding Trillium **44**,
      113

O

Orange Hawkweed **62**,
      107
Orchid, Purple Fringed
      103, 114
Oxeye Daisy 37, 39,
      43, 106

P

parsley family **24–5**,
      104
Parsnip
   Cow **30**, 104
   Water **31**, 104
   Wild 25, **49**, 104
Pasque Flower 11, **98**,
      116
Pasture Rose 79, **80**,
      116
pea family **96–7**, 111
Pearly Everlasting **32**,
      106
Pennsylvania
      Smartweed **70**,
      115
Peppermint **71**, 113
Perfoliate Bellwort **67**,
      110
Periwinkle 11,
      99, 105
Phlox, Wild Blue **99**,
      115
Pineapple Weed **50**,
      106
Pink, Fire 11,
      79, 110

Plantain
   Common 12, **33**,
      114
   Water 104
Poison Hemlock **30**,
      104
Poppy, California 11,
      14, **59**, 114
Prairie Clover, Purple
      89, 111
Prairie Coneflower **63**,
      107
Prairie Smoke 14, **73**,
      116
Prickly Pear Cactus **61**,
      109
Purple Coneflower 10,
      84, 107
Purple Fringed Orchid
      103, 114
Purple Loosestrife **92**,
      113
Purple Prairie Clover
      89, 111

Q, R

Queen Anne's Lace 31
Rabbit-foot Clover **74**,
      111
Ragweed, Common
      19, 106
Ragwort, Golden **55**,
      107
Rattlesnake Root 40
Red Clover **74**, 75, 111
Rose
   Carolina 80
   Pasture 79, **80**, 116
Rough Blazing Star **95**,
      107
Rough-fruited
      Cinquefoil **52**, 116

S

St. John's Wort,
      Common **60**, 112
seeds 14
shapes, plants **9**, 11, 12

Sharp-lobed Hepatica 26, **34**, 115
Shepherd's Purse **29**, 76, 77, 109
Skunk Cabbage 22, **23**, 105
Smartweed, Pennsylvania **70**, 115
Snakeroot, Black **30**, 104
Solomon's Seal 20, **21**, 105
Spiderwort 11, **98**, 110
Spotted Coralroot 11, **78**, 114
Spotted Knapweed **85**, 107
Spotted Touch-me-not **61**, 108
Spreading Dogbane **70**, 105
Spring Beauty **78**, 115
Spurge, Leafy 16, **20**, 110
stems 8
Stinging Nettle 9, **18**, 117
Strawberry, Common 9, **35**, 116
Sunflower, Common **64**, 107
sunflower family **38–9**, 106–8
Swamp Milkweed **71**, 105
Sweet Clover
    White **29**, 111
    Yellow **49**, 111
Sweet Joe-Pye Weed **72**, 107

T

Tall Buttercup 59
Tansy, Common **54**, 107
Teasel **95**, 110
Thistle
    Bull 10, 39, **85**, 107
    Canada **78**, 107

Toadflax, Butter-and-eggs **62**, 116
Toadshade Trillium **79**, 113
Toothwort, Cutleaf **36**, 77, 109
Touch-me-not, Spotted **61**, 108
Trefoil, Birdsfoot 11, 14, **52**, 97, 111
Trillium
    Large-Flowered **44**, 113
    Nodding 44, 113
    Toadshade **79**, 113
Tufted Vetch 91

V

Valerian **72**, 117
    Greek **93**, 115
Vervain, Blue 90
Vetch
    Cow 11, **91**, 111
    Crown 73, **75**, 97, 111
    Tufted 91
Violet
    Birdsfoot **99**, 117
    Common Blue 91, **92**, 117
    Downy Yellow **53**, 117
Viper's Bugloss **93**, 108
Virginia Bluebells 11, **100**, 108
Virginia Waterleaf **93**, 112

W

Water Parsnip **31**, 104
Water Plantain 104
Waterleaf, Virginia **93**, 112
White Clover **36**, 111
White Lettuce **40**, 106
White Sweet Clover **29**, 111
Wild Bergamot 56, **81**, 113

Wild Blue Phlox **99**, 115
Wild Carrot 11, 24, **31**, 104
Wild Cucumber 14, **33**, 110
Wild Four O'Clock **73**, 113
Wild Geranium 10, **83**, 112
Wild Ginger **79**, 105
Wild Lettuce **51**, 106
Wild Lupine **89**, 96, 111
Wild Parsnip 25, **49**, 104
Wintercress **48**, 109
Wood Anemone **41**, 115
Wood Sorrel, Yellow 9, 10, **53**, 114

Y

Yarrow **32**, 106
Yellow Flag 14, 46, **60**, 112
Yellow Goatsbeard **62**, 107
Yellow Sweet Clover **49**, 111
Yellow Wood Sorrel 9, 10, **53**, 114
Yucca **45**, 106

Acknowledgments

Dorling Kindersley would like to thank David Burnie for providing the text for the introduction and the features section.

The publisher would like to thank the following for their kind permission to reproduce their photographs:

(Key: a-above; b-below/bottom; c-center; f-far; l-left; r-right; t-top)

1 Dudley Edmondson. 4 Dudley Edmondson: (b). **8 Dudley Edmondson:** (c, br). **9 Corbis:** Patrick Johns (cr). **Dorling Kindersley:** Neil Fletcher (bl). **Dudley Edmondson:** (tl, tr, cl). **Rough Guides:** (br). **10 Dudley Edmondson:** (t, cr). **11 Corbis:** Minden Pictures / Tim Fitzharris (tl). **Dudley Edmondson:** (tr, c, cr). **Science Photo Library:** Hal Horwitz (cra). **12 Dorling Kindersley:** Neil Fletcher (bc). **14 Alamy Images:** Duncan Shaw (br). **Corbis:** Ocean (cb). **Dorling Kindersley:** Neil Fletcher (bl). **Dudley Edmondson:** (tl, tr). **Getty Images:** altrendo nature (bc); Radius Images (clb); Design Pics / Jack Goldfarb (crb). **15 Rough Guides. 16–17 Alamy Images:** kpzfoto. **19 Dudley Edmondson:** (t, b). **20 Dudley Edmondson:** (c). **22 Dudley Edmondson:** (c). **23 Lance Merry. 24 Alamy Images:** Earl Walker (c). **28 Dudley Edmondson:** (tr). **30 Dudley Edmondson:** (b, t). **31 Dudley Edmondson:** (c, b). **32 Dudley Edmondson:** (c, b). **33 Dudley Edmondson:** (b). **34 Dudley Edmondson:** (t). **35 Corbis:** Scott T. Smith (t). **Dudley Edmondson:** (c, b). **36 Dudley Edmondson:** (b). **37 Dudley Edmondson:** (b). **39 Dudley Edmondson:** (t). **40 Dudley Edmondson:** (t, c, b).**41 Dudley Edmondson:** (t, b). **42 Dudley Edmondson:** (t, c, b). **43 Dudley Edmondson:** (b). **44 Dudley Edmondson:** (t, b). **48 Dudley Edmondson:** (b). **50 Dudley Edmondson:** (b) **51 Alamy Images:** Nature Photographers Ltd (br). **Dudley Edmondson:** (t).**52 Dudley Edmondson:** (t, c). **53 Dudley Edmondson:** (t, c, b). **55 Dudley Edmondson:** (t, b). **56 Getty Images:** Frank Krahmer (c). **58 Dudley Edmondson:** (t). **61 Corbis:** DLILLC (br). **Dudley Edmondson:** (tc, tl). **62 Dudley Edmondson:** (c, b). **63 Dudley Edmondson:** (b). **64 Dudley Edmondson:** (b). **65 Dudley Edmondson:** (b). **66 Dudley Edmondson:** (t, b). **67 Alamy Images:** Trevor Chriss (b). **Getty Images:** Garden Picture Library / Joshua McCullough (t). **68–69 Rough Guides. 70 Dudley Edmondson:** (t, b). **71 Dudley Edmondson:** (t, c). **72 Alamy Images:** Custom Life Science Images (t). **73 Dudley Edmondson:** (t, b). **74 Dudley Edmondson:** (b). **75 Dudley Edmondson:** (b). **76 Corbis:** Frank Lane Picture Agency / Winfred Wisniewski (c). **77 Dudley Edmondson:** (tc). **Dorling Kindersley:** Neil Fletcher (clb). **78 Corbis:** National Geographic Society / Bates Littlehales (b). **Dudley Edmondson:** (t). **Science Photo Library:** Hal Horwitz (c). **79 Dudley Edmondson:** (t, c, b). **80 Dudley Edmondson:** (t, c, b). **82 Dudley Edmondson:** (t, b). **83 Dudley Edmondson:** (t, c, b). **85 Dudley Edmondson:** (b). **88 Dudley Edmondson:** (t, b). **89 Alamy Images:** Radius Images (t). **Dudley Edmondson:** (c). **90 Dudley Edmondson:** (b). **92 Mark Stensaas:** (c). **93 Alamy Images:** Allan Munsie (c). **Dudley Edmondson:** (t). **94 Dudley Edmondson:** (c). **95 Dudley Edmondson:** (t). **96 Corbis:** Steve Terrill (c). **98 Corbis:** Minden Pictures / Tim Fitzharris (c). **Dudley Edmondson:** (b). **99 Alamy Images:** Michael McMurrough (t). **Dudley Edmondson:** (b).**100 Dudley Edmondson:** (t, b). **101 Dudley Edmondson:** (t, b). **102 Dudley Edmondson:** (t, b). **103 Dudley Edmondson:** (t). **104 Dudley Edmondson:** (t, c, clb, crb, bl). **105 Dudley Edmondson:** (t, ca, c, br, bl). **Lance Merry:** (cra). **106 Alamy Images:** Nature Photographers Ltd (br). **Dudley Edmondson:** (tr, ca, cra, c, cr, clb, cb, crb, bl). **107 Alamy Images:** Custom Life Science Images (clb). **Dudley Edmondson:** (tl, tc, tr, cla, ca, cra, c, cr, crb, crb, bc, br). **108 Dudley Edmondson:** (tc, tr, cla, cra, cb, bc). **109 Corbis:** DLILLC (ca). **Dudley Edmondson:** (tr, cla, c, cr, cb). **110 Alamy Images:** Trevor Chriss (cla). **Corbis:** Minden Pictures / Tim Fitzharris (cra). **Dudley Edmondson:** (tl, cl, cr, cb, bc). **Getty Images:** Garden Picture Library / Joshua McCullough (tr). **111 Alamy Images:** Radius Images (cb). **Dudley Edmondson:** (cra, cr, c, clb, crb). **112 Alamy Images:** Allan Munsie (cra). **Dudley Edmondson:** (tr, cla, crb). **113 Dudley Edmondson:** (tr, cra, cl, c, cr, bl, br). **114 Dudley Edmondson:** (tl, cra, c, cb). **Science Photo Library:** Hal Horwitz (ca). **115 Corbis:** National Geographic Society/Bates Littlehales (clb). **Dudley Edmondson:** (tr, tc, cra, cr, crb, cb, bc, br). **116 Dudley Edmondson:** (cla, tr, cra, cl, c, cr, cb, bc). **117 Alamy Images:** Michael McMurrough (cl). **Dudley Edmondson:** (cra, c). **Mark Stensaas:** (cr).

All other images © Dorling Kindersley
For further information see: **www.dkimages.com**